The Aging Brain: Navigating Changes and Optimizing Performance

Nora

TABLE OF CONTENTS

CHAPTER ONE: INTRODUCTION

1.1 Background

Normal aging may lead to changes in brain structure and function. Consequently, neuroimaging studies have demonstrated that the prefrontal cortex (PFC) is highly susceptible to deterioration (Cabeza & Dennis, 2013). The PFC is responsible for mediating executive functions that are involved in everyday tasks such as walking or multitasking (Agbangla et al., 2017). More specifically, executive functions encompass higher order cognitive processes such as processing speed, neural inhibition and working memory which are essential components of attention, coordination and planning (Diamond, 2013; Miyake et al., 2000). Executive function deficits have been identified in older adults such that a redistribution of prefrontal resources may be necessary to mitigate the demands of a complex task. Reports in the literature have identified two main processes, neural efficiency (Causse et al., 2017; Grady, 2012) and neural compensation (Cabeza et al., 2018; Reuter-Lorenz & Park, 2014), which may account for the differences in brain activation between older and younger adults. Neural efficiency has been demonstrated when task-related performance is maintained or improved without the need for increased brain activation (Causse et al., 2017). In comparison, compensation mechanisms typically involve increasing brain activation but do not guarantee improved performance (Cabeza et al., 2018).

Performance and brain activity have been assessed using the dual-task paradigm (Pashler, 1994). When examining mobility, it can be used to assess the costs of walking and performing a cognitive task simultaneously (Fraser et al., 2016; Woollacott & Shumway-Cook, 2002; Yogev et al., 2008). Since both tasks compete for executive control, performance deficits in one or both tasks are commonly observed. Functional near-infrared spectroscopy (fNIRS) has been used to measure prefrontal hemodynamic changes since it can tolerate motion and participant mobility (Pinti et al., 2018). Findings from several reports revealed that older adults demonstrated a compensatory upregulation of metabolic resources in PFC compared to younger adults and, therefore, greater recruitment of executive processes with advancing age (Fraser et al., 2016; Holtzer et al., 2011; Mirelman et al., 2017). However, performance measures such as response time (Deary & Der, 2005; Vaportzis et al., 2013), accuracy (Srygley et al., 2009; Vermeij et al., 2012a), and gait speed (Smith et al., 2016; Yogev et al., 2008) prove to be more variable despite the increase in brain activation.

Alternatively, older adults may preserve neural efficiency processes from younger adulthood and adapt to increasing cognitive demands by automatizing gait (Clark, 2015). Dual-tasks facilitate an automatic locomotor control strategy when the secondary task can draw conscious attention away from walking (Wulf et al., 2001). This strategy is efficient in that it allows gait to be processed using minimal conscious control (Bernstein, 1967). Therefore, when dual-task demands exceed cognitive processing capacity, shifting from controlled (i.e., compensation) to automatic processing (i.e., greater efficiency) diminishes the demand placed on the PFC (Clark, 2015). Compensatory resources may, instead, be used to supplement motor control when there is a loss of automaticity.

The neural correlates of dual-task walking during complex cognitive tasks remains poorly understood. There is no consensus in the literature as to how older adults balance controlled and automatic processing to maintain task performance. The variability in hemodynamic response and performance seen in dual-task walk studies may be driven by an unjust comparison between dual-tasks of unequal demand (Patel et al., 2014). As such, examining processing speed, neural inhibition, and working memory cognitive tasks may help reveal how the cognitive demands of each one contributes to dual-task walking. This will equally provide a better understanding of the shared pathways between cognitive and motor control including the balance between controlled and automatic processing during dual-task walking.

1.2 Purpose

This study is part of a larger project that examined static and dynamic measurements of balance and gait in older and younger adults. The goal of this component was to collect and examine the dynamic data including the neural and behavioural measures of dual-task walking in older adults. More specifically, the purpose was to gain a better understanding of how older adults mitigate simultaneous cognitive and motor demands during dual-tasks of increasing difficulty. Based on the work of Reuter-Lorenz and Park (2014), older adults may exhibit a compensatory redistribution of prefrontal metabolic resources when task demands approach cognitive capacity. Compared to younger adults, prefrontal activation is more widespread and bilaterally distributed during cognitively demanding tasks (Reuter-Lorenz & Park, 2014). However, to avoid exceeding processing capacity, older adults may revert to automatizing gait to free up controlled processing resources in the PFC (Clark, 2015). Therefore, the aims of this study are two-fold. Firstly, to independently analyze the hemodynamic response changes ($\Delta HbO2$ and ΔHbR) and behaviour (response time, accuracy and gait speed) of older adults across processing speed, neural inhibition and working memory tasks. The second aim was to determine whether neural and behavioural measures were correlated with one another such that changes in brain activation were associated with changes in performance outcomes during the dual-tasks.

1.3 Hypotheses

1) *Neural*: In line with the neuroimaging literature, an interaction is expected between task (single-, dual-tasks) and cognitive demand (processing speed, neural inhibition, two working memory tasks), such that older adults will exhibit increased $\Delta HbO2$ and ΔHbR

across dual-task cognitive difficulty levels in comparison to single-task due to PFC recruitment for greater processing control.

2) *Behaviour*: At the lowest level of cognitive difficulty (i.e., processing speed), performance will be maintained between single- and dual-tasks. However, an interaction between task and cognitive difficulty is expected for each successive level of demand including slower response times and gait speed, and declines in accuracy performance.

3) *Correlation between neural and behaviour*: Older adults will demonstrate a relationship between cerebral oxygenation and behaviour such that increased prefrontal activation may be associated with decreased performance during the dual-tasks.

CHAPTER TWO: LITERATURE REVIEW

2.1 Executive functions and aging

The PFC plays a fundamental role in executive functioning by integrating information from various brain regions to produce a goal-oriented behaviour. More specifically, executive functions encompass higher order cognitive processes that control more basic ones to help an individual adapt to new or complex environments (Diamond, 2013). For example, everyday tasks such as walking involve attention, motor planning and the integration of sensorimotor information (Yogev et al., 2008). Therefore, the cumulative demands of walking involve executively controlled processes that implicate prefrontal brain regions. It has been suggested that executive functions decline with advancing age due to the generalized slowing of processing speed (Salthouse, 1996). However, alternate interpretations propose the examination of inhibition and working memory, as deficits in these domains are equally reported with age (Baddeley, 1986; West, 1996). The mechanisms by which executive functions operate to maintain cognitive abilities remain relatively unknown but examining the interactions between them may provide a more comprehensive overview of cognitive decline.

Processing speed accounts for the time it takes to execute a cognitive operation (Salthouse, 1996). Therefore, when processing speed is too slow, it minimizes the cognitive resources that are available to process information (Albinet et al., 2012; West, 1996). There are different ways to evaluate processing speed including paper and pencil tests such as the Trail Making Test Part A (Wechsler, 1981) and response time tasks (Vaportzis et al., 2013). These cognitive tests are designed to examine the speed in which older adults can react to a stimulus and formulate an appropriate response. However, processing speed deficits have been attributed to changes in brain structure and function. In particular, frontal regions such as the PFC are highly susceptible to deterioration through the reduction of grey matter volume and white matter integrity (Cabeza &

Dennis, 2013; Raz & Rodrigue, 2006). Reports in the neuroimaging literature suggest that these age-associated changes may lead to a reduced availability of prefrontal cerebral blood flow (CBF) and, therefore, differences in brain activation and performance between older and younger adults (Agbangla et al., 2017; Bertsch et al., 2009). Brain activation can be inferred from CBF due to a process known as neurovascular coupling (Sorond et al., 2011). In theory, a neural stimulus drives a series of cellular processes that require an increased metabolic demand of oxygen. In turn, there is an increased amount of oxygen that is delivered to the activated region of the brain. Interestingly, Rosso et al., (2017) used a reaction time task which can be used to assess processing speed. The authors reported that older adults exhibited greater prefrontal brain activation and significantly slower response times than younger adults (Rosso et al., 2017). These findings demonstrate that there may be a reorganization of cognitive resources to compensate for changes in the PFC with age.

Neural inhibition and processing speed are interrelated in that inhibition involves the ability to quickly distinguish between and then react to relevant versus irrelevant stimuli. In general, it is believed that older adults have trouble selectively inhibiting information which may affect the way that they allocate their attention towards task execution (Broadbent, 1958). For example, Hsieh et al., (2015) demonstrated that older adults were more prone to distractors compared to younger adults when differentiating between vowels and consonants in the alphabet. Similarly, the Stroop Test is frequently used measure to assess inhibition as it involves processing speed, attention and even elements of automaticity (Stroop, 1935). Briefly, this test requires participants to read aloud the name of a colour that is written in a different colour than the word itself. The colour of the word is said to be an irrelevant source of interference that must be ignored prior to responding with the written word's name. Other inhibition tests, such as the go/no-go task, soon evolved from this design whereby individuals are required to respond to a "go" stimulus and withhold their response from a "no-go" stimulus (Potvin-Desrochers et al., 2017).

Working memory proves to be even more complex than processing speed and inhibition as it involves storing and manipulating recently encoded information (Baddeley, 1986). Although it is known to be regulated under a limited processing capacity, holding onto specific information requires the selective inhibition of irrelevant stimuli, which may be more challenging for older adults (Diamond, 2013). Vermeij et al., (2012) demonstrated that older adults performed worse than younger adults during tasks of increasing working memory load. Working memory has equally been evaluated using paper and pencil tests such as the Trail Making Test Part B where participants have to connect alternating letters and numbers in ascending order (Wechsler, 1981). Other studies have incorporated an n-back task such that participants have to listen to a sequence of stimuli and then respond with the information they heard n stimuli back (Fraser et al., 2016).

2.2 Attention and the dual-task paradigm

The role of attention is to selectively inhibit irrelevant stimuli to allow for a greater focus on relevant information that helps to execute a task (Broadbent, 1958). Researchers have, therefore, employed the dual-task paradigm to assess the relative attention costs or attention priority dedicated to two simultaneous tasks (Pashler, 1994). Dual-tasks exacerbate attention capacity by forcing task switching and control over interference between two competing tasks. Similarly, overlapping processes may inhibit one another such that performance on one or both tasks may suffer (Pashler, 1994).

In contrast, a secondary task may promote automatic processing (Kal et al., 2015; Wulf et al., 2001). As stipulated in the "constrained action hypothesis," the allocation of attention between two simultaneous tasks may contribute to the automatic regulation of a motor task (Wulf, 2013;

Wulf et al., 2001). When a postural task is performed alone, attention may be directed towards an internal focus. In other words, it may lead to focusing on the movement itself, which interferes with automatic control processes. However, when a secondary task is introduced in addition to a postural control task, attention may be diverted towards an external point of focus (i.e., the secondary task). In turn, automatic processes remain unconstrained and can produce efficient movements with minimal interference. This may be applied to dual-task walking whereby the cognitive task acts as the external focus to promote walking automaticity. However, dual-tasks also involve more complex cognitive processes that contribute to diverting attention from the motor task.

Early studies evaluating dual-task walking have compared performance during normal walking and walking while talking in older adults (Lundin-Olsson et al., 1997). Findings demonstrated that some older adults could not walk and talk simultaneously because of the competing demands of both tasks. The dual-task paradigm has continued to evolve over time and has now been employed across a variety of study protocols and task types (Pelicioni et al., 2019). Studies aiming to assess gait typically employ a cognitive-motor dual-task. However, there is no consensus on protocol design or task selection which may account for inconsistent findings across studies (Beurskens & Bock, 2012; Kahya et al., 2019; Pelicioni et al., 2019). The most commonly reported cognitive tasks in the literature include verbal fluency (Hawkins et al., 2018; Holtzer et al., 2015; Verghese et al., 2017) and counting backwards (Al-Yahya et al., 2011; Mirelman et al., 2017). Although each one is meant to evaluate executive functioning, the results may not be directly comparable. For example, Holtzer et al., (2015) and Verghese et al., (2017) conducted studies that required participants to recite alternating letters of the alphabet beginning with the letters A or B. In comparison, Clark et al., (2014) and Hawkins et al., (2018) tasked participants with listing words beginning with a predetermined letter of the alphabet. Both tasks involve similar cognitive processes yet the differences between reciting the alphabet and searching for words impact dual-task performance.

2.3 Cognitive task difficulty manipulation

Dual-task studies report inconsistent findings as to whether brain activation and performance should increase, decrease or stay the same between single- and dual-tasks in older adults (Kahya et al., 2019; Pelicioni et al., 2019; Smith et al., 2017). When comparing these studies, what is often overlooked is the relative difficulty of the different types of tasks (Patel et al., 2014). In other words, less is known about whether executive functioning declines because of a general slowing of processing speed or whether executive function domains differentially contribute to cognitive decline. Mirelman et al., (2017) compared brain activation in older and younger adults and demonstrated that greater executive control was primarily observed as task demands increased. In that study, normal walking elicited greater prefrontal activation than quiet standing in older adults. Similarly, normal walking while serially subtracting three resulted in even greater activation. However, compared to the younger adults who only exhibited increased brain activation during the serial subtraction task, the magnitude of prefrontal activation change during that task was approximately the same. These findings suggest that rather than interpreting the extent in which prefrontal activation changes, it may be equally meaningful to understand which cognitive task demand levels elicit a significant change in prefrontal activation (Mirelman et al., 2017).

Gait performance may also be altered by increasing the difficulty of a cognitive task. Srygley et al., (2009) compared normal walking to walking while serially subtracting three or seven from a predetermined starting number. Findings from this study demonstrated that older adults slowed their gait speed between each successive level of cognitive demand. More specifically, there was a significant decrease in gait speed between the single- and dual-task walking conditions and a further decrease between serial subtraction difficulty levels. Cognitive performance equally declined during the dual-tasks and across difficulty such that the older adults made more mistakes and completed fewer subtractions during the seven versus three conditions. These findings suggest that manipulating cognitive demands may help reveal whether older adults alter their performance to accommodate for the increasing cognitive demands. However, it is unclear whether declines in brain activation and performance are specific to a single task type or whether older adults can adapt across different executive function domains.

2.4 Automatic and executive control of walking

Both the automatic and executive control of walking merit careful consideration when examining changes in brain activation and performance in older adults. In fact, it is unlikely that each process functions independently, rather, there may be a shift based on the task demands (Clark, 2015; Schneider & Shiffrin, 1977). Evidence from the neuroimaging literature suggests that gait is controlled by indirect and direct motor control pathways (Clark, 2015; Hamacher et al., 2015). Direct motor control involves spinal reflex pathways that quickly and efficiently integrate sensorimotor information (Clark, 2015; Woollacott & Shumway-Cook, 2002). It can thus be considered more "automatic" because minimal conscious attention is required to execute a motor movement. In comparison, the indirect pathway involves integrating information via the PFC and its executive control processes (Clark, 2015; Hamacher et al., 2015). Executive processes, or those controlled by the PFC, are limited by an individual's finite attention capacity (Broadbent, 1958). Therefore, executive resources are said to be more "controlled" as more overt attention is required to execute the task.

Two complementary theories, namely neural efficiency (Causse et al., 2017; Grady, 2012) and neural compensation (Cabeza et al., 2018; Reuter-Lorenz & Park, 2014), may be able to explain age-related changes in locomotor control strategies. Firstly, neural efficiency can be represented by a decrease in prefrontal activation and an increase in performance during a demanding task (Causse et al., 2017). In other words, performance may be facilitated by the effective allocation of processing resources. A common example of neural efficiency is the automatization of a motor task such that brain activation decreases but motor performance is maintained. Wu et al., (2004) used functional magnetic resonance imaging (fMRI) to help elucidate the neural correlates of automatic motor processes. By implementing the dual-task paradigm, participants were asked to concurrently perform a sequential finger tapping movement and a letter task to compare brain activation and performance before and after training. Following extensive training, when the participants were able to achieve a high degree of accuracy on the tapping sequence with minimal interference from the digit task, it was determined that they had reached automatic control. By simultaneously measuring brain activation, it was revealed that PFC activation decreased following training. In other words, the automatized finger tapping sequence freed attentional resources which could then be used to perform the digit task. This example may be applied to the automatic control of walking such that a decrease in brain activation is expected when automatic locomotor control is achieved.

The extent in which a motor task is learnt lends itself well to automatic processing such that executing a leant task requires less overt attention (Schneider & Shiffrin, 1977; Yogev-Seligmann et al., 2012). Most healthy older adults can achieve motor task automaticity with training or practice (Seidler et al., 2010; Wollesen & Voelcker-Rehage, 2014; Wu et al., 2004). The PFC is highly involved in effortful learning due to the attentional demands associated with acquiring a new skill. When a skill is being learned, simultaneous brain regions such as the basal ganglia and associated motor circuitry begin to passively integrate new information based on the task's requirements (Dietrich & Audiffren, 2011). As such, once a task is well-learned, brain activation in the PFC will progressively decrease as less attention is required to perform the task (Poldrack, 2005; Wu et al., 2004). A shift to other brain regions that are responsible for more implicitly controlled processes can then be expected. Walking, in particular, is a well-learned task from childhood, therefore, when demands are not excessive, healthy older adults may use these automatic pathways to minimize the need for conscious attention while walking.

In contrast, aging may be associated with a loss of automaticity and greater reliance on executive control (Clark, 2015; Seidler et al., 2010). As such, some older adults may demonstrate compensatory patterns of neural activation by over-activating the PFC. Neural compensation mechanisms can be defined as a necessary increase in brain activation in order to accomplish a task (Holtzer et al., 2015; Stern, 2009). However, compared to neural efficiency, greater activation does not always correspond with enhanced performance (Cabeza et al., 2002) The Scaffolding Theory of Aging and Cognition (STAC-r) describes a life-course model of cognition that accounts for changes in brain structure and function with age (Reuter-Lorenz & Park, 2014). More specifically, it describes compensatory mechanisms such as greater and more widespread neural recruitment in older adults to counteract neural degradation. This may account for greater brain activation in older compared to younger adults when there are no differences in performance.

Neuroimaging studies examining neural compensation indicate that greater activation may (Cabeza et al., 2002; Mirelman et al., 2017) or may not (Holtzer et al., 2015; Stern, 2009) be associated with better gait performance in healthy older adults. For example, by comparing memory recall, Cabeza et al., (2002) revealed that older adults who demonstrated greater and more widespread brain activation performed better on a memory task. In comparison, Holtzer et al., (2015) compared brain activation during normal walking and walking while reciting alternating letters of the alphabet (i.e., alpha condition). The older adults demonstrated greater and more widespread activation in the PFC during the alpha condition compared to normal walking. However, compared to Cabeza et al., (2002), greater prefrontal activation did not lead to any changes in gait speed in the study conducted by Holtzer et al. (2015). These findings demonstrate that neural compensation mechanisms may not be equally effective at maintaining cognitive as compared to motor performance.

Neural efficiency or compensation and, therefore, automatic or executive control mechanisms, depend on many factors. Firstly, health status and cognition vary greatly across normal aging. When there is a known loss of automaticity, such as in the case of frail older adults or individuals with neurological disorders, greater executive control is expected during walking tasks (Holtzer et al., 2016; Maidan et al., 2016). Similarly, older adults at risk of decline who pass paper and pencil tests but demonstrate significant changes in brain activity and performance may be unable to process tasks automatically. Instead, they may invoke more top-down executive control processes (Verghese et al., 2014). However, high-functioning and cognitively healthy older adults may demonstrate intact neural networks that allow for gait to be automatically processed (Cabeza et al., 2002; Düzel et al., 2010; Vermeij et al., 2014). For example, Vermeij et al., (2014) examined a sample of high- and low-functioning older adults and found that compared to the high-functioning group, low-functioning older adults demonstrated increased brain activation and worse performance on a working memory task. Neuropsychological tests may further be used to distinguish high- and low-functioning groups and may help justify neural resource allocation (Holtzer et al., 2006). This has been examined by Holtzer et al., (2006) who demonstrated that neuropsychological performance on speed and executive attention, memory and verbal IQ tests may predict gait speed variability in older adults.

Lastly, one factor that may delineate both automatic and executive control is the finite availability of prefrontal resources. Older adults may upregulate prefrontal resources up until they reach their maximum activation capacity. Once this limit is reached, there is an inevitable brain activation drop-off (Cabeza et al., 2018). To avoid experiencing a drop-off, there may be a redistribution of metabolic resources that prioritizes the brain regions that are most critical to task execution and away from the less essential regions (Bruya, 2010). For example, cognitive resources that help maintain gait are especially important for older adults to ensure walking stability. In comparison, dedicating processing resources to accomplish a simultaneous cognitive task may be less important (Woollacott & Shumway-Cook, 2002). Similarly, acute bouts of exercise such as walking may induce a reduction of prefrontal activation in favour of greater activation in motor regions. This has been outlined in the theory of hypofrontality which suggests that exercise can induce changes in metabolic resource allocation (Dietrich, 2003). The PFC, in particular, demonstrates decreased activity during exercise due to the reduced need for higher order cognitive processing and an increased need for motor performance (Dietrich & Audiffren, 2011). Therefore, walking may induce a similar pattern in older adults such that brain regions involved in walking stability are prioritized over cognitive performance (Yogev-Seligmann et al., 2012).

2.5 Functional near-infrared spectroscopy (fNIRS)

Dual-task walking studies have been restricted by neuroimaging techniques such as fMRI in that they are sensitive to motion and require participants to remain immobilized (Bürki et al., 2017). More recently, functional near-infrared spectroscopy (fNIRS) has emerged as a better alternative to measure prefrontal activation during walking tasks because of its wearable nature, cost and ability to measure brain activation when participants are in motion (Quaresima & Ferrari, 2019). In addition, it is non-invasive and can be used on people of all ages allowing for a wide variety of study designs (Pinti et al., 2018). FNIRS measures changes in cerebral oxygenation ($\Delta HbO2$) and deoxygenation (ΔHbR) based on neurovascular coupling. In response to a neural stimulus, a series of vascular events are initiated to mitigate an increased metabolic demand of oxygen. This includes greater CBF and oxygenated hemoglobin (HbO2) delivery to the activated region of the brain and decreased deoxygenated hemoglobin (HbR). Therefore, the change in metabolic demand of oxygen and CBF can be coupled to generate a neurophysiological marker to detect changes in cerebral oxygenation (Al-Yahya et al., 2016; Sorond et al., 2011).

Continuous wave fNIRS detects changes in hemodynamic response by emitting a constant stream of near-infrared light (NIR; 700-1000 nm) through pairs optodes that either emit or detect the NIR light. The optodes may be placed along the surface of the scalp based on pre-existing templates such as the modified International EEG 10-20 system (Herwig et al., 2003). This ensures that the device is placed consistently throughout all measurements and that the region of interest is being examined. NIR light can penetrate biological tissue, therefore, the distinct absorption spectra of HbO2 and HbR demonstrate different attenuation properties when they are measured at different wavelengths (Jöbsis, 1977). Following this, several pre-processing steps are required to convert light intensities to changes in cerebral oxygenation. This is outlined in the modified Beer-Lambert law which accounts for the difference between the incident and transmitted light and is proportional to the changes in cerebral oxygenation concentration (Kocsis et al., 2006). Another aspect that is incorporated into the modified Beer-Lambert law is the differential pathlength factor (DPF) which accounts for any signal loss due to light scattering (Scholkmann & Wolf, 2013). This is especially important when examining different population groups as it accounts for the variances in head thickness and vasculature that are associated with aging.

Due to its portability and motion tolerance, fNIRS can be used to simultaneously examine the changes in brain activation and performance in older adults. Evidence from studies evaluating walking while reciting alternating letters of the alphabet (Holtzer et al., 2015; Verghese et al., 2017) and n-back working memory tasks (Fraser et al., 2016; Lövdén et al., 2008; Vermeij et al., 2012a) have demonstrated that older adults exhibit increased cerebral oxygenation between single- and dual-tasks. These findings have been attributed to the greater executive demands associated with dual-tasking in which older adults utilize greater controlled processes to mitigate the cognitive-motor demands. In comparison, Beurskens et al., (2014) demonstrated decreased cerebral oxygenation in older adults between single- and dual-tasks. This may be due to the complex visual dual-task in which the older adults had to check-off boxes while walking. Compared to cognitive-auditory dual-tasks, the older adults in this study were deprived of sensorimotor information which may have caused greater neural activation in regions outside the PFC. Certain fNIRS studies have identified activation in motor regions including the premotor (Lu et al., 2015) and supplemental motor area (Harada et al., 2009; Lu et al., 2015; Miyai et al., 2001) which may be more important during dual-task walking to maintain balance stability.

2.6 Performance measures of behaviour

Cognitive and motor performance measures can be used to track different behavioural outcomes following a neural stimulus (Al-Yahya et al., 2016). Further analyses may even determine whether a change in performance is correlated with a change in brain activation (Holtzer et al., 2015). Various measures of behaviour have been used in the literature including gait speed, response time and accuracy to examine whether older adults modify their performance in response to dual-tasking. Gait speed is one of the most commonly reported performance measures as it offers a quantitative assessment of walking performance (Kahya et al., 2019; Smith et al., 2016). Older adults typically demonstrate a decrease in gait speed between single- and dual-tasks due to the competing demands of cognitive-motor tasks (Hausdorff et al., 2008; Holtzer et al., 2016; Verghese et al., 2017). However, it may also be a strategy for older adults to optimize postural control. As outlined in the "posture first hypothesis," there may exist a hierarchy of attentional prioritization to ensure sufficient attention is allocated to postural control (Shumway-Cook et al., 1997). Li et al., (2001) employed a cognitive-motor dual-task that required older adults to memorize a list of words while walking. Findings revealed that compared to younger adults, older adults performed better on the walking task as compared to word memorization. As such, older adults may have inadvertently chosen to prioritize gait over cognitive performance to ensure safe ambulation (Shumway-Cook et al., 1997; Yogev-Seligmann et al., 2010). In addition, Li et al., (2001) demonstrated that when given an external aid to improve dual-task performance, older adults used it to prioritize gait compared to younger adults who used it to prioritize cognitive performance.

Instruction prioritization (i.e., focus on walking or focus on the cognitive task) may equally alter gait performance during dual-tasks. Studies examining prioritization compared the effects explicitly instructing older adults to focus on gait or the cognitive task compared to equally prioritizing both tasks (Verghese et al., 2007; Yogev-Seligmann et al., 2010). Verghese et al., (2007) and Yogev-Seligmann et al., (2010) offer converging evidence such that gait prioritization instructions led to increased gait speed while cognitive task prioritization resulted in decreased gait speed. However, when instructed to equally prioritize walking and the cognitive task, older adults did not demonstrate slower gait speed (Verghese et al., 2007). These findings suggest that when older adults consciously prioritize gait, they can maintain their gait performance. However, in line with the posture first hypothesis, when given a choice between cognitive or motor task performance, older adults may innately prioritize gait performance (Shumway-Cook et al., 1997).

Furthermore, response time measurements are commonly used to assess cognitive performance during dual-task studies. Studies have demonstrated that older adults respond slower during dual- versus single tasks and with increasing task demands (Lajoie et al., 1996; Tang & Wakayama, 2011). Lajoie et al., (1996) demonstrated that older adults exhibited slower responses during a wide versus narrow postural stance. These findings may be attributed to the increased attentional demands that are associated with the instability of a narrow postural stance. In addition, Tang et al., (2011) replicated these findings by examining response time during dual-task walking. Using an auditory processing speed task, older adults demonstrated prolonged response times during dual- versus single-tasks.

A complementary measure of response time is the accuracy of the response. For example, Vaportzis et al., (2013) evaluated response time and accuracy during a choice reaction time task that was paired with an easy or hard digit recall task. As expected, findings revealed that older adults responded slower than younger adults during the harder digit task. Interestingly, however, accuracy remained similar between both groups across both levels of difficulty. The interaction between speed and accuracy are two concepts that are not always mutually exclusive. Studies have demonstrated that there may be a speed-accuracy trade-off in which older adults subconsciously prioritize responding more accurately than responding quickly (Forstmann et al., 2011; Rabbitt, 1979). The first reason for this relates to the general slowing of processing speed that requires older adults to take more time to accumulate task-relevant information before reacting or responding (Salthouse, 1996). A second perspective is that a characteristic of older adults is to try to avoid making mistakes at all costs (Rabbitt, 1979). Despite this, accuracy findings prove to be more variable than response time across studies in the literature. Numerous studies have reported that older adults may (Brustio et al., 2017; Fraser et al., 2016) or may not (Rosso et al., 2017; Vaportzis et al., 2013) make more errors with increasing cognitive demands.

2.7 Summary

The PFC plays an essential role in mediating executive functions including the attentional demands associated with dual-tasking (Fraser et al., 2016). However, older adults may experience executive function deficits due to the deterioration of the PFC with advancing age (Agbangla et al., 2017). The dual-task paradigm may, therefore, be used to assess executive functions by examining how older adults mitigate the demands of two simultaneous tasks. For example, excessively demanding tasks may result in different patterns of brain activation and performance as compared to minimally demanding tasks. Reuter-Lorenz and Park (2014) attribute these changes to compensation mechanisms such that greater brain activation is required by older adults to accomplish dual-tasks in a similar manner as younger adults. In contrast, some older adults may exhibit more efficient processing mechanisms when neural networks remain intact from younger adulthood (Causse et al., 2017). This would be represented by a decrease in prefrontal brain activation while performance is maintained during a complex task (Causse et al., 2017; Grady, 2012).

Numerous studies have examined dual-task walking in older adults, however, there is no consensus as to whether brain activation as and performance should increase, decrease or stay the same between single- and dual-task walking (Pelicioni et al., 2019). In addition, many studies do not account for the effect of cognitive task difficulty on cognitive and motor performance (Patel et al., 2014). Therefore, manipulating cognitive task difficulty across different executive function domains may reveal whether executive functions universally decline or whether differences arise with increasing cognitive demands.

CHAPTER THREE: MANUSCRIPT

Hemodynamic and Behavioural Changes in Older Adults During Cognitively Demanding Dual-Tasks

1. INTRODUCTION

Declines in cognition are more common as people age and have been supported by studies examining changes in brain activation between older and younger adults (Fraser et al., 2016; C. Grady, 2000; Holtzer et al., 2011). Neuroimaging findings suggest that compensatory neural mechanisms exist to counteract decline and to allow for the maintenance of cognition over time (Cabeza et al., 2018; Reuter-Lorenz & Park, 2014). One example is the revised Scaffolding Theory of Aging and Cognition (STAC-r) which outlines compensatory scaffolding as an adaptive measure for older adults to generate and recruit additional neural resources to replace those that have deteriorated over time (Reuter-Lorenz & Park, 2014). This theory can account for greater brain activation in older versus younger adults when behavioural measures are similar between both groups (Cabeza et al., 2018).

Behavioural measures of performance such as gait speed have also been used to evaluate cognition (Al-Yahya et al., 2011). Early research has demonstrated that some older adults are unable to walk and talk at the same time and those that stopped walking to talk were more prone to falling (Lundin-Olsson et al., 1997). While walking alone did not lead to any gait changes, slowing down or stopping may be an involuntary strategy exhibited by older adults to prioritize gait and ensure safe ambulation (Holtzer et al., 2016; Shumway-Cook et al., 1997). Alternatively, higher functioning and cognitively healthy older adults may resemble younger adults in that they exhibit an automatic locomotor control strategy to manage walking and talking simultaneously (i.e., dual-tasking) (Bernstein, 1967). Automatic control is efficient in that steady state walking can be achieved under minimal conscious attention thereby freeing up executive resources for a secondary task (Clark, 2015; Poldrack, 2005). However, studies have demonstrated that greater task difficulty may lead to a loss of automaticity and greater reliance on the prefrontal cortex (PFC) due to the attentional demands associated with maintaining gait performance (Clark, 2015; Holtzer et al., 2015). This is known as the executive control of walking, which operates under a limited processing capacity, but may be recruited when dual-tasks require greater executive resources (Beurskens & Bock, 2012; Yogev et al., 2008).

The PFC is responsible for mediating complex cognitive processes namely planning, attention and coordination which are involved in everyday tasks such as walking or dual-tasking (Cabeza et al., 2018). In fact, the dual-task paradigm measures changes in executive functioning by comparing brain activation and performance between single- and dual-tasks (Pashler, 1994). Reviews in the literature demonstrate inconsistent findings as to whether prefrontal activation and behaviour should increase, decrease or stay the same between single- and dual-tasks (Kahya et al., 2019; Pelicioni et al., 2019). This may be due to diverse cognitive tasks such as verbal fluency (Hawkins et al., 2018; Holtzer et al., 2015; Verghese et al., 2017) and counting backwards (Al-Yahya et al., 2011; Mirelman et al., 2017) which differentially engage executive functions and the PFC. Therefore, it may important to account for differences in cognitive task difficulty between studies (Patel et al., 2014). One approach to mitigate this concern is a study design that targets the examination of executive functioning across multiple task difficulties. This may also allow for the identification of easier cognitive tasks that are not sensitive enough or do not challenge older adults sufficiently to detect changes in single- versus dual-tasks. More specifically, this may reveal whether executive control is only evoked under greater cognitive demands and whether STAC-r compensatory mechanisms are efficient enough to preserve performance.

In order to simultaneously examine the neural and behavioural mechanisms underlying executive functioning, functional near-infrared spectroscopy (fNIRS) can be used to monitor cerebral oxygenation ($\Delta HbO2$) and deoxygenation (ΔHbR) changes in the PFC. FNIRS is advantageous over other functional neuroimaging techniques most notably for its non-invasive and portable nature that doesn't limit an individual's mobility (Pinti et al., 2018). In its application to walking, it tolerates motion artifacts better than other techniques and can be used on people of all ages with no adverse health consequences (Pinti et al., 2018). FNIRS exploits the transient nature of biological tissue to near-infrared light as well as the distinct absorption spectra of oxygenated (HbO2) and deoxygenated (HbR) hemoglobin in the near-infrared region (Quaresima & Ferrari, 2019). In theory, the PFC requires an influx of HbO2 and efflux of HbR as cognitive demands increase. Therefore, during dual-tasks, the increased cerebral blood flow and metabolic demand of oxygen can be coupled in a process known as neurovascular coupling (Quaresima & Ferrari, 2019). This process can then serve as a neurophysiological marker for fNIRS to detect changes in cerebral oxygenation during dual-task walking studies (Al-Yahya et al., 2016; Sorond et al., 2011).

Furthermore, various behavioural measures can be used to quantify the shift from performance maintenance to decline. Firstly, gait speed is a commonly used measure to assess locomotor control (Hausdorff et al., 2008; Smith et al., 2016; Yogev et al., 2008). Studies have demonstrated a strong relationship between poor executive functioning and slower gait speed especially during dual-tasks involving a challenging locomotor component (Hawkins et al., 2018; Maidan et al., 2016; Mirelman et al., 2017). This is in line with the executive processing of gait which is recruited when tasks are unlearned or too challenging to be automatically processed (Clark, 2015). Other behavioural measures such as response time and accuracy have been reported in the literature but with greater variability across different task types and difficulty levels. For example, by using a cognitive-auditory response time task, Rosso et al., (2017) found slower response times in dual- compared to single-tasks but no differences in accuracy (Rosso et al., 2017). In contrast, studies examining neural inhibition and working memory have demonstrated that performance declines in older adults in dual- compared to single-tasks (Fraser et al., 2016; Hsieh et al., 2016). This may be due to the complex processing steps involved in discerning relevant from irrelevant stimuli during an inhibition task and temporarily storing and manipulating information during a working memory task both of which are particularly challenging for older adults (Baddeley, 1986; Hsieh et al., 2016). As such, the present study is unique in that it will evaluate various executive processing domains by manipulating cognitive demands according to an easy processing speed task, a medium level neural inhibition task and two difficult working memory tasks.

The purpose of this study was to examine how older adults mitigate the demands of dual-tasking through changes in brain activation and behaviour. The first aim was to determine the changes in cerebral oxygenation ($\Delta HbO2$ and ΔHbR) using fNIRS and performance (gait speed, cognitive response time and accuracy) in single- versus dual-tasks and across four levels of cognitive task difficulty. Greater cerebral oxygenation changes were expected during the dual-tasks in comparison to single-tasks and these changes were expected to increase with each successive difficulty level. Performance was expected to decrease between single- and dual-tasks with the most significant change occurring during the working memory tasks. The second aim was to correlate cerebral oxygenation and behaviour to determine whether increased brain activation would be associated with poorer performance during the dual-tasks. Therefore, understanding neural and behavioural changes in healthy older adults may help reveal whether declines are only associated with specific executive function domains.

2. METHODS

2.1 Participants

Twenty healthy older adults ($M = 71.8$ years, $SD = 6.4$ years, 10 females) were recruited from community centres across Ottawa, Canada. Participant eligibility was determined using a phone screening (**Table 1**) whereby participants were included if they were right-handed according to the Edinburgh Handedness Inventory (Oldfield, 1971) and did not have a diagnosed hearing impairment or hearing aid. Participants also had to be comfortable walking 15 meters without assistance and without neuromuscular or physical complaints that could affect walking (i.e., severe arthritis). Cognitive status was determined using the Montreal Cognitive Assessment (MoCA) where participants were required to score ≥ 26 to ensure that they were cognitively healthy (Nasreddine et al., 2005). This study was ethically approved by the University of Ottawa Research Ethics Board and all participants provided written informed consent before participating in the study.

2.2 FNIRS equipment

Participants were fitted with a wearable OctaMon fNIRS device (Octamon, Artinis, The Netherlands) to measure prefrontal $\Delta HbO2$ and ΔHbR. The distance between the nasion and inion was measured for each participant to ensure the fNIRS device was placed along the PFC according to the modified International EEG 10-20 system (Herwig et al., 2003). The OctaMon uses continuous wave near-infrared spectroscopy, which measures near-infrared light absorption at two distinct wavelengths (760 and 850 nm). This device also uses eight light emitting diode (LED) channels and two detectors with an interoptode distance of 35 mm (**Figure 1**).

2.3 Experimental protocol

Participants were presented with four runs in a randomized order each evaluating one of four levels of cognitive demands. A run was comprised of 12 counterbalanced blocks with an equal number of single cognitive (SC), single motor (SM) and dual-tasks (DT) blocks (**Figure 2**). In the SC condition, participants performed the cognitive task while standing and staring straight ahead at a target. The SM block had participants walk without a cognitive task at their self-selected pace along a 10 m walkway. During the DT condition, participants were asked to perform both the cognitive and motor task simultaneously and were instructed to pay equal attention to both tasks. To gain a better understanding of the subjective emphasis dedicated to the dual-tasks, participants were asked to report how much attention (out of a possible 100%) they attributed to the cognitive and motor task following the DT blocks. Each 33 s block was preceded by a 10 s baseline of quiet standing and was followed by a 15 s rest period to allow the hemodynamic response to revert to the baseline in between blocks. Throughout the experiment, participants were given breaks as needed and upon request.

2.4 Cognitive task difficulty levels

E-Prime software (version 2.0) was used to create different cognitive task sequences. The experimenter delivered all instructions to the participants using a microphone which could be heard through wireless headphones worn by the participant. Four cognitive-auditory tasks: simple reaction time (SRT), go/no-go (GNG), n-back (NBK) and double number sequence (DNS), were chosen from previous work in our labs, to represent processing speed (SRT), neural inhibition (GNG) and working memory tasks (NBK and DNS) (Fraser et al., 2016; St-Amant et al., 2020). During a short practice session, participants familiarized themselves with the cognitive tasks until they were able to correctly respond to 70% of the SC stimuli. The SRT task represented the simplest cognitive demand and had participants respond to a random sequence of beeps (2850 Hz at 99 dB) by saying the word "top" as quickly as possible following each stimulus. GNG was the medium level task and had participants listen to both high- (2850 Hz at 99 dB) and low-pitched (970 Hz at 95 dB) beeps but only respond "top" to the high-pitched beeps. The next level task was the NBK and had participants listen to a continuous sequence of single-digit numbers (1-9) and respond with the number they heard two numbers back. Lastly, the DNS task represented the highest cognitive demand and had participants listen to a sequence of three-digit numbers. At the end of the block, they reported the total number of times they heard two target digits within the entire sequence (Richer et al., 2017). Two working memory tasks (NBK and DNS) were chosen because working memory is the executive domain known to be most affected by cognitive aging (Baddeley, 1986).

2.5 Behavioural measures

Three behavioural measures were chosen to evaluate performance differences between single and dual-tasks as well as across cognitive task difficulty. The first measure, gait speed (m/s), was calculated by dividing the distance the participants walked by the fixed duration of the block. Response times (s) were recorded using a voice recorder and imported into Audacity (version 2.3.1) to measure the time from stimulus onset until the participant's response. Response times were recorded during the SRT, GNG and NBK difficulty levels. A response time was not measured during the DNS condition because it is a non-verbal working memory task that has participants withhold their response until the end of the block. Finally, experimenters calculated accuracy scores (% correct) for correct responses to the cognitive tasks. In the SRT difficulty level, correct responses were recorded when the participant responded to a beep by saying the word "top" while incorrect responses were noted when the participant did not respond to a beep. Correct responses in the GNG condition were calculated when the participant correctly responded to the high- rather than the low-pitched beep. Errors were noted when participants either missed the high-beep or responded to the low beep. During the NBK, correct responses involved participants correctly responding with the number they heard two numbers back. Errors were given when participants responded with the incorrect number or did not respond at all. Finally, correct responses in the

DNS were calculated based on the participant's final tally of each target digit compared to the total possible correct responses.

2.6 Test battery

Following the experiment, participants were asked to complete a battery of neuropsychological and physical tests. The purpose of these tests was to ensure good cognitive and physical function, low fear of falling and no depression which may influence study outcomes. The neuropsychological tests included the Montreal Cognitive Assessment (MoCA) (Nasreddine et al., 2005), Digit Forward and Backward (Wechsler, 1981), Digit Symbol Substitution Test (Wechsler, 1981) and Trail Making Test (TMT) Part A and B (Strauss et al., 2006). The MoCA is a screening tool used to assess cognitive impairment. Individuals who score ≥ 26 out of 30 reflect healthy cognition (Nasreddine et al., 2005). Digit Forward and Backward are used to assess working memory and points were awarded for correctly repeating a growing list of numbers in either the forward or reverse direction. The Digit Symbol Substitution Test measures processing speed as individuals fill-in as many symbols as possible within 90 s based on a key provided at the top of the worksheet. The Trail Making Tests are timed tests (s) used to measure task switching and executive functioning. It is divided into two parts whereby Part A has participants draw lines connecting 25 ascending numbers while Part B has participants draw lines alternating between ascending numbers and letters. A shorter time to complete these tests indicates better performance. Furthermore, physical status and fear of falling were assessed using the Short Physical Performance Battery (SPPB) and the Falls Efficacy Scale-International (FES-I), respectively. The SPPB measures lower extremity functioning in older adults and is scored out of 12, where 12 is equivalent to no deficits in functioning (Guralnik et al., 1994). FES-I uses a 4-point Likert scale to assess an individual's fear of falling (Delbaere et al., 2010). It is scored out of 64 whereby a higher score indicates a greater fear of falling. The Geriatric Depression Scale was also used to assess depression in older adults as it is known to have effects on the PFC (Yesavage & Sheikh, 1986). It is scored out of 30 and a lower score within the range of 0-9 indicates no depression.

2.7 Data processing of fNIRS signal

Neural data was collected in Oxysoft (version 3.0.97.1) and sampled at a frequency of 10 Hz. After visually inspecting the signal quality, the Modified Beer-Lambert law was applied to the raw HbO2 and HbR intensities using a differential pathlength factor set to 6.61 for all older adults (Scholkmann & Wolf, 2013). The concentrations were then preprocessed offline using a custom MATLAB (version R2018a) script. The script eliminated motion artifacts by removing outliers that were 2.5 SD from the mean and replaced them with a zero value. Additionally, in line with the literature, a Butterworth bandpass filter set between 0.01-0.14 Hz was used to reduce physiological noise (heartbeat and breathing) within the signal (Holtzer et al., 2011; Mirelman et al., 2017; Verghese et al., 2017). An average ΔHbO2 and ΔHbR value was then calculated in μM for each task (SC, SM, DT) and each difficulty level (SRT, GNG, NBK, DNS) from the changes in signal between the baseline and active conditions.

2.8 Statistical analyses

Differences in cerebral oxygenation (ΔHbO2 and ΔHbR) were assessed using 2x4 repeated measures ANOVAs whereby task (SC/SM vs. DT) and difficulty (SRT, GNG, NBK, DNS) main effects and interactions were tested.

Assessments of behavioural response time were tested with a 2x3 repeated measures ANOVA to measure the interaction between task (SC, DT) and difficulty (SRT, GNG, NBK). Note that the DNS task had participants respond at the end of the block, therefore, no response time was calculated. Significant differences in gait speed and accuracy were evaluated with 2x4 repeated measures ANOVAs to measure the interaction between task (SC/SM vs. DT) and difficulty (SRT, GNG, NBK, DNS).

A one-way ANOVA was conducted on the subjective emphasis responses to test whether there were significant differences between how much attention the participants dedicated to walking versus the cognitive tasks across each difficulty level (SRT, GNG, NBK, DNS).

For all repeated measures ANOVAs, if Mauchly's Test of Sphericity was violated, a Greenhouse-Geisser p-value was reported. In addition, Bonferroni post hoc analysis was used to determine the location of significance where statistical significance was set at $p < 0.05$. Means (M) and standard deviations (SD) are reported in the results and when a distinction between difficulty levels is needed, the difficulty level is identified in subscript (i.e., M_{SRT} = Mean value for SRT difficulty level). Means and standard deviations were calculated for all participant demographics and neuropsychological assessments.

No significant differences were observed in terms of cerebral oxygenation between channels or hemispheres (*p-values* > 0.05). Therefore, brain activation was analyzed across the whole PFC by averaging the concentration output from each channel. In addition, we verified if there were significant changes in cerebral oxygenation within task (e.g., the four SM blocks in SRT) and there were no significant differences (*p-values* > 0.90). As such, an average of each task type was calculated for analyses.

A Pearson correlation was used to examine the relationship between cerebral oxygenation ($\Delta HbO2$ and ΔHbR) and performance (gait speed, response time and accuracy) during the dual-tasks.

3. RESULTS

3.1 Neural: Changes in cerebral oxygenation

A significant main effect of task on $\Delta HbO2$ was observed $F (1,19) = 4.5, p = 0.047, \eta^2 = 0.191$ (**Figure 3a**). A post hoc analysis revealed that $\Delta HbO2$ significantly decreased ($p = 0.047$) from SM ($M = 0.078$ µM, $SD = 0.026$ µM) to DT ($M = 0.028$ µM, $SD = 0.029$ µM). There was also a main effect of task on ΔHbR $F (1, 19) = 4.8, p = 0.040, \eta^2 = 0.203$ (**Figure 3b**). The post hoc analysis indicated that ΔHbR significantly decreased ($p = 0.040$) from SM ($M = 0.064$ µM, $SD = 0.021$ µM) to DT ($M = 0.021$ µM, $SD = 0.024$ µM). A normal distribution of the $\Delta HbO2$ and ΔHbR signals over the course of SM and DT blocks has been depicted in **Figure 4.** There were no significant interactions between task (SC, DT) and difficulty (SRT, GNG, NBK, DNS) for $\Delta HbO2$ ($p = 0.400$) and ΔHbR ($p = 0.412$) or main effects of task ($\Delta HbO2; p = 0.200$, HbR; $p = 0.169$) and difficulty ($\Delta HbO2; p = 0.414$, $\Delta HbR; p = 0.476$).

3.2 Behavioural: Changes in response time, accuracy and gait speed

Response time (ms) increased across increasing levels of difficulty whereby SRT < GNG < NBK ($M_{SRT} = 394$ ms, $SD_{SRT} = 86.3$ ms; $M_{GNG} = 559$ ms, $SD_{GNG} = 116$ ms; $M_{NBK} = 605$ ms, $SD_{NBK} = 206$ ms). This was demonstrated by a main effect of difficulty on response time $F (2, 38) = 16.0, p < 0.001, \eta^2 = 0.456$ (**Figure 5**). Post hoc analysis indicated that SRT response times were significantly faster than the GNG and NBK conditions ($p < 0.001$).

Analyses revealed a main effect of task on accuracy $F (1, 19) = 5.7, p = 0.028, \eta^2 = 0.230$ (**Figure 6a**). Post hoc tests revealed that SC ($M = 89.3\%, SD = 13.5\%$) was significantly more accurate than DT ($M = 86.9\%, SD = 14.4\%, p < 0.001$). There was also a main effect of difficulty $F (3, 57) = 16.2, p < 0.001, \eta^2 = 0.460$, whereby accuracy decreased as the cognitive tasks became more difficult ($M_{SRT} = 100\%, SD_{SRT} = 0.0\%; M_{GNG} = 92.0\%, SD_{GNG} = 17.0\%; M_{NBK} = 80.6\%,$

SD_{NBK} = 15.0%; M_{DNS} = 79.7%, SD_{DNS} = 4.83%) (**Figure 6b**). Post hoc tests revealed that responses in SRT were significantly more accurate than GNG (p = 0.038), NBK (p < 0.001) and DNS (p < 0.001). In addition, GNG was more accurate than NBK (p = 0.042) and DNS (p = 0.002), however, NBK and DNS were not significantly different (p = 0.740).

An interaction effect between task (SM, DT) and difficulty (SRT, GNG, NBK, DNS) was observed for gait speed, F (3, 57) = 2.2, p = 0.014, η^2 = 0.169 (**Figure 7**). Post hoc analyses indicated that during the most difficult cognitive task, the DNS, there was a significant decrease (p = 0.003) in gait speed between SM (M = 1.11 m/s, SD = 0.38 m/s) and DT (M = 1.09 m/s, SD = 0.38 m/s). There were no significant differences between single- and dual-task gait speed during the SRT (p = 0.772), GNG (p = 0.706) and NBK (p = 0.379) cognitive tasks.

The ANOVA on subjective emphasis revealed a significant decrease in the attention dedicated to walking across cognitive task difficulty F (3, 57) = 14.8, p < 0.001, η^2 = 0.438. The participants reported focusing less on walking with each successive difficulty level (M_{SRT} = 39.1%, SD_{SRT} = 18.0%; M_{GNG} = 31.4%, SD_{GNG} = 15.2%; M_{NBK} = 22.6%, SD_{NBK} = 17.4%; M_{DNS} = 18.9%, SD_{DNS} = 19.8%). Post hoc analyses revealed that participants focused significantly less on walking during the NBK (p < 0.001) and DNS (p = 0.001) compared to the SRT and significantly less in the NBK (p = 0.038) and DNS (p = 0.017) compared to the GNG. There were no significant differences between the SRT and GNG (p = 0.056) and the NBK and DNS (p = 1.000).

3.3 Correlation between cerebral oxygenation and behaviour

There were no significant correlations between cerebral oxygenation ($\Delta HbO2$; *p-values* > 0.081) and deoxygenation (ΔHbR; *p-values* > 0.068) and behaviour (response time, response accuracy, gait speed) during the dual-tasks.

4. DISCUSSION

The current study applied fNIRS imaging to assess whether older adults demonstrated changes in prefrontal cerebral oxygenation and behaviour while walking with cognitive tasks of increasing difficulty. The aims of this study were two-fold. Firstly, to analyze neural and behavioural measures to better understand neural compensation mechanisms during dual-tasks of different difficulty levels. Secondly, to determine whether there was a correlation between neural and behavioural outcomes such that increases PFC activation may be associated with better performance, or vice versa, in older adults. In doing so, this may reveal how older adults mitigate their attention capacity through prefrontal executive involvement or adopt compensatory neural strategies to meet the demands of difficult dual-tasks.

4.1 Neural

According to our initial hypothesis, $\Delta HbO2$ was expected to increase from single- to dual-tasks based on the principles of STAC-r (Reuter-Lorenz & Park, 2014). This prediction was based on the neuroimaging literature which suggests that older adults exhibit more widespread and bilateral activation in the PFC during dual- versus single-tasks and, therefore, greater dependency on executive control compared to younger adults (Fraser et al., 2016; Holtzer et al., 2011). Contrary to this expectation, this study demonstrated a significant decrease in $\Delta HbO2$ and ΔHbR between walking alone (i.e., single-task) and walking with a cognitive task (i.e., dual-task). These findings are in line with several reports that observed a decrease of prefrontal cerebral oxygenation and an alternative strategy to mitigate the demands of dual-task walking (Beurskens et al., 2014; Pelicioni et al., 2019). One possibility is an automatic locomotor control strategy which would be beneficial in dual-task situations to minimize interference with other controlled processes (Clark, 2015; Schneider & Shiffrin, 1977). The PFC's contributions to walking include managing the attentional demands and motor planning associated with safe and efficient displacement (Clark, 2015; Yogev et al., 2008). However, executive resources are limited and may be reorganized depending on task demands. Studies have shown that decreased PFC activation is associated with automatically controlled tasks and walking, in particular, is amenable to automation because it is well-learned (Dietrich & Audiffren, 2011; Wu et al., 2004). Therefore, increased prefrontal activation may only be observed in individuals who show a loss of automaticity such as in people with neurological disorders or frail older adults (Beurskens & Bock, 2012; Holtzer et al., 2016; Maidan et al., 2016; Woollacott & Shumway-Cook, 2002). Based on the data presented in **Table 2** the older adults in this study demonstrated high scores in cognitive function, walk speed (i.e., > 1 m/s) and no frailty, amongst other factors, which are typically associated with decreased executive functioning. These measures suggest that our participant group was high functioning and could rely on an automatic locomotor strategy to free up cognitive resources in the PFC.

Participants were also asked to subjectively rate how much attention they paid towards the cognitive versus walking task. Their responses reflected an automatic control strategy in that they reported focusing <39% on walking during all the cognitive tasks. The cognitive tasks may have also served as an external focus which has been known to facilitate automatic processing (Bernstein, 1967; Poldrack, 2005). This has been outlined in the "constrained action hypothesis" which suggest that focusing on the outcome of a movement (i.e., external focus), rather than the movement itself (i.e., internal focus), minimizes interference with other consciously controlled tasks (Wulf, 2013; Wulf et al., 2001). Similarly, diverting attention away from a postural task (i.e., to a cognitive task) even when cognitive demands are low may provide an external focus to improve motor performance (Huxhold et al., 2006). As such, compared to walking alone, responding to the various stimuli during the dual-tasks may have helped draw attention away from walking and allowed for greater stability without greater recruitment of the PFC. Conversely, in the absence of a cognitive task, attention could be drawn to both internal and external sources thereby engaging greater executive control.

Healthy individuals inherently shift between automatic and executive control strategies to mitigate cognitive demands (Clark, 2015; Yogev et al., 2008). However, studies have also demonstrated age-related decreases in cerebral blood flow (CBF) to the PFC due to changes in brain structure (Bertsch et al., 2009). The reorganization of locomotor control pathways and a reduction of CBF with age may, therefore, contribute to an overall reduced availability of prefrontal oxygenation. Dietrich's (2003) theory of hypofrontality suggests that there is a redistribution of metabolic resources from prefrontal brain regions to motor regions during tasks such as walking due to the complex integration of sensory, motor and autonomic processes. In other words, the brain is limited by a finite supply of metabolic resources that must be strategically allocated based on the most critical demands (Dietrich, 2003). Taken together with automaticity, hypofrontality may cause a downregulation of metabolic resources in the PFC which can be redistributed to other brain regions to supplement motor control. Regions outside the PFC could not be measured within the scope of this study, however, studies have shown heightened brain activation in motor areas such as the premotor (Lu et al., 2015) and supplemental motor area (Harada et al., 2009; Lu et al., 2015; Miyai et al., 2001) during dual-task walking. These brain regions should be further examined simultaneously with the PFC to determine whether a decrease in prefrontal cerebral oxygenation from single- to dual-task corresponds with changes in motor regions when walking more automatically.

We must also acknowledge certain study parameters including the (i) cognitive and (ii) motor tasks that differentiate this study from others in the literature. (i) Cognitive tasks: Verbal fluency (Hawkins et al., 2018; Holtzer et al., 2015; Verghese et al., 2017) and counting backwards (Al-Yahya et al., 2016; Mirelman et al., 2017) are the most commonly used tasks in dual-task studies that demonstrate increased or no change in cerebral oxygenation between single- and dual-tasks (Pelicioni et al., 2019). Our study used processing speed, neural inhibition and working memory tasks which continuously prompted responses and engaged participants based on a random sequence of stimuli. This differs from verbal fluency and counting tasks in that participants were not provided with a starting cue (i.e., a letter or number) after which they could respond at their own pace. The external focus of the cognitive tasks and unpredictable pattern of stimuli may have helped recruit automatic control pathways by ensuring that the full duration of the task was attention-demanding (Beck et al., 2018; Bernstein, 1967). (ii) Motor task: Walking trajectories vary significantly across studies due to equipment and space constraints. As evidenced by studies examining obstacle negotiation, the interruption of steady state walking caused increased PFC activation and may equally impede automaticity (Hawkins et al., 2018; Holtzer et al., 2016; Maidan et al., 2016). Our study provided participants with a 10 m pathway to maximize straight-line walking which is considerably longer than studies examining gait along electronic walkways (Hernandez et al., 2016; Holtzer et al., 2011; Mirelman et al., 2017; Verghese et al., 2017). Therefore, our walking task provided longer stretches of steady state walking and a greater opportunity to automatize gait than studies using shorter walkways.

Lastly, in addition to the $\Delta HbO2$ decrease, there was also a decrease in ΔHbR between the single- and dual-tasks. ΔHbR is a reliable measure of neural activation but is less commonly reported in the literature. This is due to its low signal amplitude making significant changes between baseline and task conditions more difficult to obtain (Leff et al., 2011). The low signal amplitude also means that HbR is less likely to be contaminated with physiological artifacts and also results in a lower signal to noise ratio (Leff et al., 2011). As such, capturing a significant HbR change that mirrors the HbO2 findings further supports a decrease in brain activation between single- and dual-tasks.

4.2 Behavioural

Examining gait speed in older adults alongside behavioural measures such as response time and accuracy may offer insights into the cognitive-motor interactions underlying dual-task walking. Gait speed changes in older adults have been well documented in the literature such that increasing attentional demands while walking may affect walking performance (Hausdorff et al., 2008; Smith et al., 2016; Yogev et al., 2008). Findings from the present study partially support this in that gait speed decreased but only during the most difficult cognitive task. Gait speed maintenance across the first three levels of task difficulty may be explained by an automatic locomotor control strategy, as described in the neural findings. However, this strategy may not have been sufficient to mitigate the demands of the DNS dual-task. As suggested in the "posture first hypothesis," older adults subconsciously prioritize gait over cognitive performance to ensure safe ambulation (Holtzer et al., 2016; Shumway-Cook et al., 1997; Yogev-Seligmann et al., 2010). Slowing gait speed may, therefore, be a combination of prioritization and compensation strategies to ensure older adults can function safely under complex task demands. It is worth noting that older adults commonly decrease their gait speed <1.0 m/s during dual-tasks which is also a cut-off used to identify individuals who are at a greater risk of falls (Hollman, McDade, et al., 2011; Kyrdalen et al., 2019; Smith et al., 2016; Verghese et al., 2017). When the older adults in this study decreased their gait speed during the most difficult task, it still remained on average >1.0 m/s. This may further indicate the physical status of the participants which could have an impact on performance as compared to other studies in the literature (Dupuy et al., 2015; Holtzer et al., 2015).

Decreased response time and accuracy performance may also be a consequence of gait prioritization. Our findings demonstrated increased response times from the easiest to the most demanding task. More specifically, the response times in the SRT task were significantly faster than the GNG and NBK tasks. However, the GNG and NBK tasks were not significantly different from one another. This was expected in that compared to the SRT task, the GNG and NBK tasks involved more complex processing steps. For example, the simple reaction time task required a response after each stimulus whereas the GNG task forced the older adults to first discriminate between a "go" and "no-go" stimulus before responding (Hsieh et al., 2016). Similarly, the NBK working memory task involved maintaining and updating information before responding to the stimuli (Al-Yahya et al., 2011). Based on these findings, more complex processing steps require more processing capacity. This was evident during the more difficult tasks as the older adults slowed their response times significantly during the inhibition and the working memory tasks compared to the SRT task. Further, the older adults responded less accurately as the difficulty level increased. However, there were no differences between the working memory tasks. These findings support our difficulty manipulation such that participants were most accurate during the processing speed task and least accurate during the working memory tasks.

In line with the literature, increasing task difficulty was expected to result in lower accuracy (Fraser et al., 2016; Srygley et al., 2009; Vermeij et al., 2012b). Interestingly, participants maintained their accuracy >80% throughout all the dual-tasks. This suggests that a high level of performance is achievable with increasing cognitive demands when cognitive resources are allocated effectively. However, participants were less accurate during the dual- versus single-tasks. This has been demonstrated in the literature whereby participants make more errors during dual-tasks due to the competing demands of performing two tasks simultaneously (Brustio et al., 2017; Srygley et al., 2009).

4.3 Correlation between cerebral oxygenation and behaviour

There were no significant correlations between the changes in cerebral oxygenation and behavioural performance. More specifically, the changes in cerebral oxygenation across task and difficulty were not associated with gait speed, response time or accuracy performance. This could be due to the small sample of older adults in this study. However, interpreting neural and behavioural findings together revealed that the redistribution of metabolic resources in the PFC may have contributed to insignificant differences in gait speed across the first three levels of task difficulty. The same cannot be said for response time and accuracy performance in which decreased cerebral oxygenation in the PFC did not result in behavioural gains. Future studies should examine automaticity and neural efficiency across task difficulty in regions outside the PFC as certain regions of interest may increase or decrease activity with the maintenance and decline of different performance measures. Follow-up studies should be conducted to determine how this impacts cognition in the long-term. This may equally reveal whether individuals exhibiting decrements in behaviour due to neural inefficiency may be at a greater risk of cognitive decline.

4.4 Limitations

Gait parameters were only quantified using gait speed. Gait speed is commonly used in the literature because it is easily collected in clinical settings, it requires minimal equipment and is a good indicator of motor performance in older adults (Holtzer et al., 2015). However, other measures that capture gait variability including stride length or stride time could complement gait speed measures and may provide greater insight into subtle changes in dual-task performances, different age groups, and different clinical populations. In addition, the choice of fNIRS device limited our data acquisition to the PFC (Octamon, Artinis, The Netherlands). This device facilitated setup and caused minimal discomfort for the participants, however, we can only speculate as to which other brain regions were involved in dual-tasking and the potential executive-automatic processing shift in walking with increasing difficulty. Despite this, fNIRS has a high temporal resolution compared to other techniques such as fMRI and is a reliable tool for measuring cerebral oxygenation in the PFC (Pinti et al., 2018).

5. CONCLUSION

Executive functions are known to decline with age and can significantly affect the way older adults divide their attention between two simultaneous tasks. Many older adults adapt to these changes by using compensatory neural strategies to accomplish tasks exceeding their cognitive capacity. The neural findings of this study suggest that an automatic locomotor control strategy can decrease the recruitment of executive resources in the PFC during dual- versus single-tasks. Behaviourally, this allowed for gait speed maintenance until the most difficult working memory task after which older adults slowed down to mitigate the cognitive task demands. Consequently, prioritizing gait led to slower response times and worse response accuracy across task difficulty.

Findings from this study helped reveal the PFC's role in allocating cognitive resources during processing speed, neural inhibition and working memory tasks while walking. Future studies can develop an even better understanding of this relationship as neuroimaging becomes more portable, more extensive (i.e., covering the entire brain) and adaptable to different environments. In particular, assessing dual-task walking in real-life situations such as crossing the street while talking on the phone may generate more novel approaches to understanding executive and controlled processes within the scope of cognitive aging.

6. REFERENCES

Al-Yahya, E., Dawes, H., Smith, L., Dennis, A., Howells, K., & Cockburn, J. (2011). Cognitive motor interference while walking: A systematic review and meta-analysis. *Neuroscience & Biobehavioral Reviews*, *35*(3), 715–728. https://doi.org/10.1016/j.neubiorev.2010.08.008

Al-Yahya, E., Johansen-Berg, H., Kischka, U., Zarei, M., Cockburn, J., & Dawes, H. (2016). Prefrontal cortex activation while walking under dual-task conditions in stroke: A multimodal imaging study. *Neurorehabilitation and Neural Repair*, *30*(6), 591–599. https://doi.org/10.1177/1545968315613864

Baddeley, A. (1986). *Working memory*. Oxford: Oxford University Press.

Beck, E., Intzandt, B., & Almeida, Q. J. (2018). Can dual task walking improve in Parkinson's disease after external focus of attention exercise? A single blind randomized controlled trial. *Neurorehabilitation and Neural Repair*, *32*(1), 18–33. https://doi.org/10.1177/1545968317746782

Bernstein, N. (1967). *The co-ordination and regulation of movements*. Oxford, New York, Pergamon Press.

Bertsch, K., Hagemann, D., Hermes, M., Walter, C., Khan, R., & Naumann, E. (2009). Resting cerebral blood flow, attention, and aging. *Brain Research*, *1267*, 77–88. https://doi.org/10.1016/j.brainres.2009.02.053

Beurskens, R., & Bock, O. (2012). Age-related deficits of dual-task walking: A review. *Neural Plasticity*, *2012*. https://doi.org/10.1155/2012/131608

Beurskens, R., Helmich, I., Rein, R., & Bock, O. (2014). Age-related changes in prefrontal activity during walking in dual-task situations: A fNIRS study. *International Journal of Psychophysiology*, *92*(3), 122–128. https://doi.org/10.1016/j.ijpsycho.2014.03.005

Brustio, P., Magistro, D., Zecca, M., Rabaglietti, E., & Liubicich, M. (2017). Age-related decrements in dual-task performance: Comparison of different mobility and cognitive tasks. A cross sectional study. *PLOS ONE*, *12*(7), e0181698. https://doi.org/10.1371/journal.pone.0181698

Cabeza, R., Albert, M., Belleville, S., Craik, F., Duarte, A., Grady, C., Lindenberger, U., Nyberg, L., Park, D., Reuter-Lorenz, P., Rugg, M., Steffener, J., & Rajah, M. (2018). Maintenance, reserve and compensation: The cognitive neuroscience of healthy ageing. *Nature Reviews Neuroscience*. https://doi.org/10.1038/s41583-018-0068-2

Clark, D. (2015). Automaticity of walking: Functional significance, mechanisms, measurement and rehabilitation strategies. *Frontiers in Human Neuroscience*, *9*. https://doi.org/10.3389/fnhum.2015.00246

Delbaere, K., Close, J., Mikolaizak, A., Sachdev, P., Brodaty, H., & Lord, S. (2010). The falls efficacy scale international (FES-I). A comprehensive longitudinal validation study. *Age and Ageing*, *39*(2), 210–216. https://doi.org/10.1093/ageing/afp225

Dietrich, A. (2003). Functional neuroanatomy of altered states of consciousness: The transient hypofrontality hypothesis. *Consciousness and Cognition*, *12*(2), 231–256. https://doi.org/10.1016/S1053-8100(02)00046-6

Dietrich, A., & Audiffren, M. (2011). The reticular-activating hypofrontality (RAH) model of acute exercise. *Neuroscience & Biobehavioral Reviews*, *35*(6), 1305–1325. https://doi.org/10.1016/j.neubiorev.2011.02.001

Dupuy, O., Gauthier, C., Fraser, S., Desjardins-Crèpeau, L., Desjardins, M., Mekary, S., Lesage, F., Hoge, R., Pouliot, P., & Bherer, L. (2015). Higher levels of cardiovascular fitness are associated with better executive function and prefrontal oxygenation in younger and older women. *Frontiers in Human Neuroscience, 9*. https://doi.org/10.3389/fnhum.2015.00066

Fraser, S., Dupuy, O., Pouliot, P., Lesage, F., & Bherer, L. (2016). Comparable cerebral oxygenation patterns in younger and older adults during dual-task walking with increasing load. *Frontiers in Aging Neuroscience, 8*. https://doi.org/10.3389/fnagi.2016.00240

Grady, C. (2000). Functional brain imaging and age-related changes in cognition. *Biological Psychology, 54*(1), 259–281. https://doi.org/10.1016/S0301-0511(00)00059-4

Guralnik, J., Simonsick, E., Ferrucci, L., Glynn, R., Berkman, L., Blazer, D., Scherr, P., & Wallace, R. (1994). A short physical performance battery assessing lower extremity function: Association with self-reported disability and prediction of mortality and nursing home admission. *Journal of Gerontology, 49*(2), M85–M94. https://doi.org/10.1093/geronj/49.2.M85

Harada, T., Miyai, I., Suzuki, M., & Kubota, K. (2009). Gait capacity affects cortical activation patterns related to speed control in the elderly. *Experimental Brain Research, 193*(3), 445–454. https://doi.org/10.1007/s00221-008-1643-y

Hausdorff, J., Schweiger, A., Herman, T., Yogev-Seligmann, G., & Giladi, N. (2008). Dual task decrements in gait among healthy older adults: contributing factors. *The Journals of Gerontology. Series A, Biological Sciences and Medical Sciences, 63*(12), 1335–1343.

Hawkins, K., Fox, E., Daly, J., Rose, D., Christou, E. A., McGuirk, T., Otzel, D., Butera, K., Chatterjee, S., & Clark, D. (2018). Prefrontal over-activation during walking in people with mobility deficits: Interpretation and functional implications. *Human Movement Science, 59*, 46–55. https://doi.org/10.1016/j.humov.2018.03.010

Hernandez, M., Holtzer, R., Chaparro, G., Jean, K., Balto, J., Sandroff, B., Izzetoglu, M., & Motl, R. (2016). Brain activation changes during locomotion in middle-aged to older adults with multiple sclerosis. *Journal of the Neurological Sciences, 370*, 277–283. https://doi.org/10.1016/j.jns.2016.10.002

Herwig, U., Satrapi, P., & Schönfeldt-Lecuona, C. (2003). Using the international 10-20 EEG system for positioning of transcranial magnetic stimulation. *Brain Topography, 16*(2), 95–99. https://doi.org/10.1023/B:BRAT.0000006333.93597.9d

Hollman, J., McDade, E., & Petersen, R. (2011). Normative spatiotemporal gait parameters in older adults. *Gait & Posture, 34*(1), 111–118. https://doi.org/10.1016/j.gaitpost.2011.03.024

Holtzer, R., Mahoney, J., Izzetoglu, M., Izzetoglu, K., Onaral, B., & Verghese, J. (2011). FNIRS study of walking and walking while talking in young and old individuals. *The Journals of Gerontology Series A: Biological Sciences and Medical Sciences, 66A*(8), 879–887. https://doi.org/10.1093/gerona/glr068

Holtzer, R., Mahoney, J., Izzetoglu, M., Wang, C., England, S., & Verghese, J. (2015). Online fronto-cortical control of simple and attention-demanding locomotion in humans. *NeuroImage, 112*, 152–159. https://doi.org/10.1016/j.neuroimage.2015.03.002

Holtzer, R., Verghese, J., Allali, G., Izzetoglu, M., Wang, C., & Mahoney, J. (2016). Neurological gait abnormalities moderate the functional brain signature of the posture first hypothesis. *Brain Topography, 29*(2), 334–343. https://doi.org/10.1007/s10548-015-0465-z

Hsieh, S., Wu, M., & Tang, C. (2016). Adaptive strategies for the elderly in inhibiting irrelevant and conflict no-go trials while performing the go/no-go task. *Frontiers in Aging Neuroscience, 7.* https://doi.org/10.3389/fnagi.2015.00243

Huxhold, O., Li, S., Schmiedek, F., & Lindenberger, U. (2006). Dual-tasking postural control: Aging and the effects of cognitive demand in conjunction with focus of attention. *Brain Research Bulletin, 69*(3), 294–305. https://doi.org/10.1016/j.brainresbull.2006.01.002

Kahya, M., Moon, S., Ranchet, M., Vukas, R., Lyons, K. E., Pahwa, R., Akinwuntan, A., & Devos, H. (2019). Brain activity during dual task gait and balance in aging and age-related neurodegenerative conditions: A systematic review. *Experimental Gerontology,* 110756. https://doi.org/10.1016/j.exger.2019.110756

Kyrdalen, I., Thingstad, P., Sandvik, L., & Ormstad, H. (2019). Associations between gait speed and well-known fall risk factors among community-dwelling older adults. *Physiotherapy Research International, 24*(1), e1743. https://doi.org/10.1002/pri.1743

Leff, D., Orihuela-Espina, F., Elwell, C., Athanasiou, T., Delpy, D., Darzi, A., & Yang, G. (2011). Assessment of the cerebral cortex during motor task behaviours in adults: A systematic review of functional near infrared spectroscopy (fNIRS) studies. *NeuroImage, 54*(4), 2922–2936. https://doi.org/10.1016/j.neuroimage.2010.10.058

Lu, C., Liu, Y., Yang, Y., Wu, Y., & Wang, R. (2015). Maintaining gait performance by cortical activation during dual-task interference: A functional near-infrared spectroscopy study. *PLoS ONE, 10*(6). https://doi.org/10.1371/journal.pone.0129390

Lundin-Olsson, L., Nyberg, L., Gustafson, Y., Himbert, D., Seknadji, P., Karila-Cohen, D., Juliard, J., & Steg, P. (1997). *"Stops walking when talking" as a predictor of falls in elderly people. 349,* 1.

Maidan, I., Nieuwhof, F., Bernad-Elazari, H., Reelick, M., Bloem, B. R., Giladi, N., Deutsch, J., Hausdorff, J., Claassen, J., & Mirelman, A. (2016). The Role of the frontal lobe in complex walking among patients with Parkinson's disease and healthy older adults: An fNIRS study. *Neurorehabilitation and Neural Repair*, *30*(10), 963–971. https://doi.org/10.1177/1545968316650426

Mirelman, A., Maidan, I., Bernad-Elazari, H., Shustack, S., Giladi, N., & Hausdorff, J. (2017). Effects of aging on prefrontal brain activation during challenging walking conditions. *Brain and Cognition*, *115*, 41–46. https://doi.org/10.1016/j.bandc.2017.04.002

Miyai, I., Tanabe, H., Sase, I., Eda, H., Oda, I., Konishi, I., Tsunazawa, Y., Suzuki, T., Yanagida, T., & Kubota, K. (2001). Cortical mapping of gait in humans: a near-infrared spectroscopic topography study. *NeuroImage*, *14*(5), 1186–1192. https://doi.org/10.1006/nimg.2001.0905

Nasreddine, Z., Phillips, N., Bédirian, V., Charbonneau, S., Whitehead, V., Collin, I., Cummings, J., & Chertkow, H. (2005). The Montreal cognitive assessment, MoCA: A brief screening tool for mild cognitive impairment. *Journal of the American Geriatrics Society*, *53*(4), 695–699. https://doi.org/10.1111/j.1532-5415.2005.53221.x

Oldfield, R. (1971). The assessment and analysis of handedness: The Edinburgh inventory. *Neuropsychologia*, *9*(1), 97–113. https://doi.org/10.1016/0028-3932(71)90067-4

Pashler, H. (1994). *Dual-task interference in simple tasks: data and theory*. *116*(2), 25. https://doi.org/10.1037/0033-2909.116.2.220

Patel, P., Lamar, M., & Bhatt, T. (2014). Effect of type of cognitive task and walking speed on cognitive-motor interference during dual-task walking. *Neuroscience*, *260*, 140–148. https://doi.org/10.1016/j.neuroscience.2013.12.016

Pelicioni, P., Tijsma, M., Lord, S., & Menant, J. (2019). Prefrontal cortical activation measured by fNIRS during walking: Effects of age, disease and secondary task. *PeerJ*, *7*, e6833. https://doi.org/10.7717/peerj.6833

Pinti, P., Tachtsidis, I., Hamilton, A., Hirsch, J., Aichelburg, C., Gilbert, S., & Burgess, P. (2018). The present and future use of functional near-infrared spectroscopy (fNIRS) for cognitive neuroscience. *Annals of the New York Academy of Sciences*, *0*(0), 1–25. https://doi.org/10.1111/nyas.13948

Poldrack, R. (2005). The neural correlates of motor skill automaticity. *Journal of Neuroscience*, *25*(22), 5356–5364. https://doi.org/10.1523/JNEUROSCI.3880-04.2005

Quaresima, V., & Ferrari, M. (2019). A mini-review on functional near-infrared spectroscopy (fNIRS): Where do we stand, and where should we go? *Photonics*, *6*(3), 87. https://doi.org/10.3390/photonics6030087

Reuter-Lorenz, P., & Park, D. (2014). How does it STAC up? Revisiting the scaffolding theory of aging and cognition. *Neuropsychology Review*, *24*(3), 355–370. https://doi.org/10.1007/s11065-014-9270-9

Richer, N., Saunders, D., Polskaia, N., & Lajoie, Y. (2017). The effects of attentional focus and cognitive tasks on postural sway may be the result of automaticity. *Gait & Posture*, *54*, 45–49. https://doi.org/10.1016/j.gaitpost.2017.02.022

Rosso, A., Cenciarini, M., Sparto, P., Loughlin, P., Furman, J., & Huppert, T. (2017). Neuroimaging of an attention demanding dual-task during dynamic postural control. *Gait & Posture*, *57*, 193–198. https://doi.org/10.1016/j.gaitpost.2017.06.013

Sala, S., Baddeley, A., Papagno, C., & Spinnler, H. (1995). Dual-task paradigm: A means to examine the central executive. *Annals of the New York Academy of Sciences*, *769*(1 Structure and), 161–172. https://doi.org/10.1111/j.1749-6632.1995.tb38137.x

Schneider, W., & Shiffrin, R. (1977). *Controlled and automatic human information processing: I. detection, search, and attention.* 66.

Scholkmann, F., Kleiser, S., Metz, A., Zimmermann, R., Mata Pavia, J., Wolf, U., & Wolf, M. (2014). A review on continuous wave functional near-infrared spectroscopy and imaging instrumentation and methodology. *NeuroImage*, *85*, 6–27. https://doi.org/10.1016/j.neuroimage.2013.05.004

Scholkmann, F., & Wolf, M. (2013). General equation for the differential pathlength factor of the frontal human head depending on wavelength and age. *Journal of Biomedical Optics*, *18*(10), 105004. https://doi.org/10.1117/1.JBO.18.10.105004

Shumway-Cook, A., Woollacott, M., Kerns, K., & Baldwin, M. (1997). The effects of two types of cognitive tasks on postural stability in older adults with and without a history of falls. *The Journals of Gerontology Series A: Biological Sciences and Medical Sciences*, *52A*(4), M232–M240. https://doi.org/10.1093/gerona/52A.4.M232

Smith, E., Cusack, T., & Blake, C. (2016). The effect of a dual task on gait speed in community dwelling older adults: A systematic review and meta-analysis. *Gait & Posture*, *44*, 250–258. https://doi.org/10.1016/j.gaitpost.2015.12.017

Sorond, F., Kiely, D., Galica, A., Moscufo, N., Serrador, J., Iloputaife, I., Egorova, S., Dell'Oglio, E., Meier, D. S., Newton, E., Milberg, W., Guttmann, C., & Lipsitz, L. (2011). Neurovascular coupling is impaired in slow walkers: The MOBILIZE Boston Study. *Annals of Neurology*, *70*(2), 213–220. https://doi.org/10.1002/ana.22433

Srygley, J., Mirelman, A., Herman, T., Giladi, N., & Hausdorff, J. (2009). When does walking alter thinking? Age and task associated findings. *Brain Research, 1253*, 92–99. https://doi.org/10.1016/j.brainres.2008.11.067

St-Amant, G., Rahman, T., Polskaia, N., Fraser, S., & Lajoie, Y. (2020). Unveilling the cerebral and sensory contributions to automatic postural control during dual-task standing. *Human Movement Science, 70*, 102587. https://doi.org/10.1016/j.humov.2020.102587

Strauss, P., Sherman, N. & Spreen, O. (2006). *A compendium of neuropsychological tests: administration, norms, and commentary*. Oxford University Press.

Verghese, J., Wang, C., Ayers, E., Izzetoglu, M., & Holtzer, R. (2017). Brain activation in high-functioning older adults and falls: Prospective cohort study. *Neurology, 88*(2), 191–197. https://doi.org/10.1212/WNL.0000000000003421

Vermeij, A., van Beek, A., Olde Rikkert, M, Claassen, J., & Kessels, R. (2012). Effects of aging on cerebral oxygenation during working-memory performance: A functional near-infrared spectroscopy study. *PLoS ONE, 7*(9), e46210. https://doi.org/10.1371/journal.pone.0046210

Wechsler, D. (1981). *Wechsler adult intelligence scale-revised (WAIS-R)*. Psychological Corporation.

Woollacott, M., & Shumway-Cook, A. (2002). Attention and the control of posture and gait: A review of an emerging area of research. *Gait & Posture, 16*(1), 1–14. https://doi.org/10.1016/S0966-6362(01)00156-4

Wu, T., Kansaku, K., & Hallett, M. (2004). How self-initiated memorized movements become automatic: A functional MRI study. *Journal of Neurophysiology, 91*(4), 1690–1698. https://doi.org/10.1152/jn.01052.2003

Wulf, G., Shea, C., & Park, J. (2001). Attention and motor performance: Preferences for and advantages of an external focus. *Research Quarterly for Exercise and Sport*, *72*(4), 335–344. https://doi.org/10.1080/02701367.2001.10608970

Yesavage, J., & Sheikh, A. (1986). Geriatric depression scale (GDS). *Clinical Gerontologist*, *5*(1–2), 165–173. https://doi.org/10.1300/J018v05n01_09

Yogev, G., Hausdorff, J., & Giladi, N. (2008). The role of executive function and attention in gait. *Movement Disorders : Official Journal of the Movement Disorder Society*, *23*(3), 329–472. https://doi.org/10.1002/mds.21720

Yogev-Seligmann, G., Hausdorff, J., & Giladi, N. (2012). Do we always prioritize balance when walking? Towards an integrated model of task prioritization. *Movement Disorders*, *27*(6), 765–770. https://doi.org/10.1002/mds.24963

7. TABLES

Table 1. Summary of participant characteristics from the phone screening (Mean ± SD).

Characteristic	n = 20
Age (years)	71.8 ± 6.4
Gender	
Male	10
Female	10
Education (years)	17.0 ± 2.4
No. of medications	1.15 ± 1.0
No. of falls while walking	0.15 ± 0.37
No. of participants who exercise more than 2x/week	19

Table 2. Mean neuropsychological and health status test scores (Mean ± SD).

Test	n = 20
MoCA (/30)	27.2 ± 1.2
Digit Forward (score/16)	10.7 ± 1.6
Digit Backward (score/14)	7.3 ± 1.9
Digit Symbol Substitution Test (# of symbols /93)	45.0 ± 9.7
TMT A (s)	37.9 ± 12.7
TMT B (s)	83.3 ± 25.6
SPPB (/12)	11.1 ± 1.6
FES-I (/64)	20.8 ± 3.6
GDS (/30)	3.2 ± 3.0

8. FIGURES

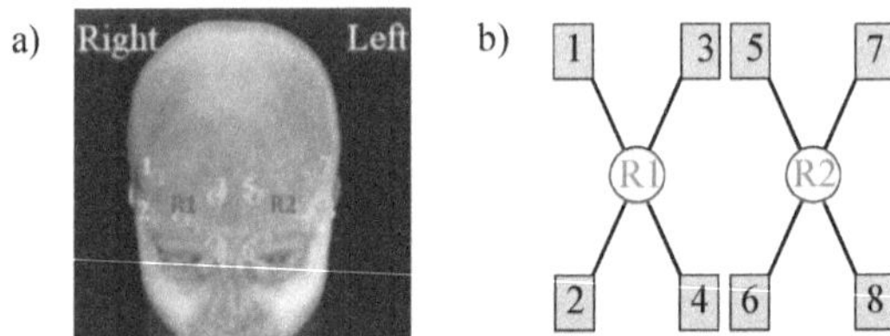

Figure 1. a) Localization of fNIRS optodes across the PFC. b) Optode template for the OctaMon fNIRS device that includes eight infrared light sources (1-8) and two detectors (R1 and R2).

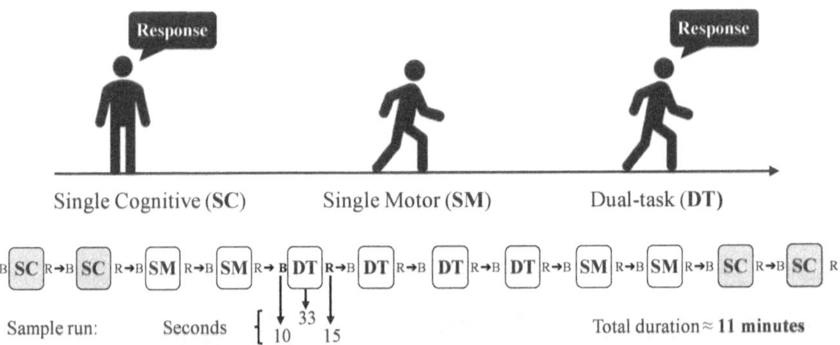

Figure 2. Description of a sample run including single cognitive (SC; responding to the cognitive task), single motor (SM; normal walking) and dual-task (DT; walking with a cognitive task) blocks. Each 33 s block is preceded by a 10 s baseline and followed by a 15 s rest period. The approximate duration of a run is 11 minutes and is repeated for each cognitive task difficulty level.

a)

b)

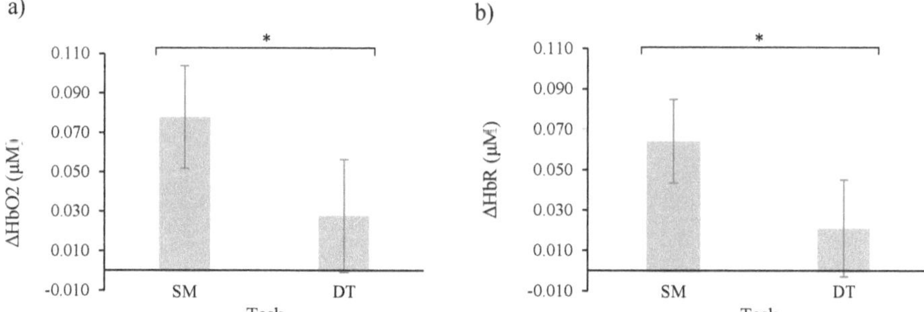

Figure 3. a) Change in prefrontal cerebral oxygenation (ΔHbO2) $F_{(1,19)} = 4.5$, $p = 0.047$, $\eta^2 = 0.191$ between single motor (SM) and dual-task (DT). There was a significant decrease in PFC activation during the dual- versus single-task ($p = 0.047$). b) Change in prefrontal cerebral deoxygenation (ΔHbR) $F_{(1, 19)} = 4.8$, $p = 0.040$, $\eta^2 = 0.203$ between single motor (SM) and dual-task (DT) blocks. Cerebral deoxygenation in the PFC significantly decreased between single- and dual-tasks ($p = 0.040$). (*) indicates significance $p < 0.05$. Error bars represent standard error of the mean.

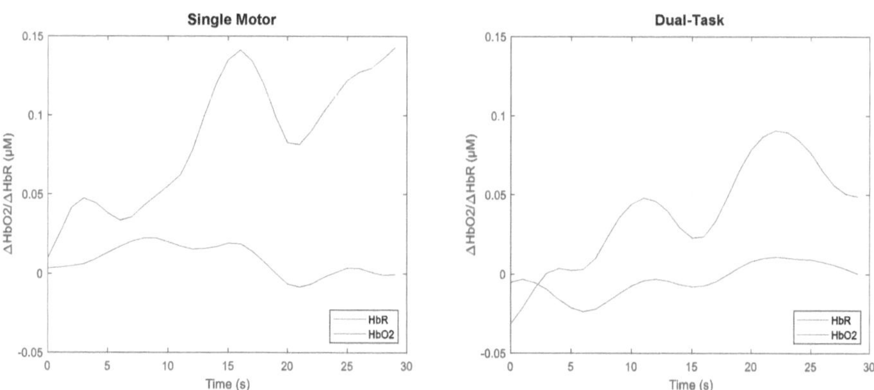

Figure 4. Mean hemodynamic response across all participants in the single motor and dual-task blocks. Neural findings demonstrated a main effect of task such that oxygenated hemoglobin (HbO2; $F_{(1,19)} = 4.5$, $p = 0.047$, $\eta^2 = 0.19$) and deoxygenated hemoglobin (HbR; $F_{(1, 19)} = 4.8$, $p = 0.040$, $\eta^2 = 0.203$) significantly decreased between single motor and dual-tasks. The blue and red lines represent HbR and HbO2, respectively.

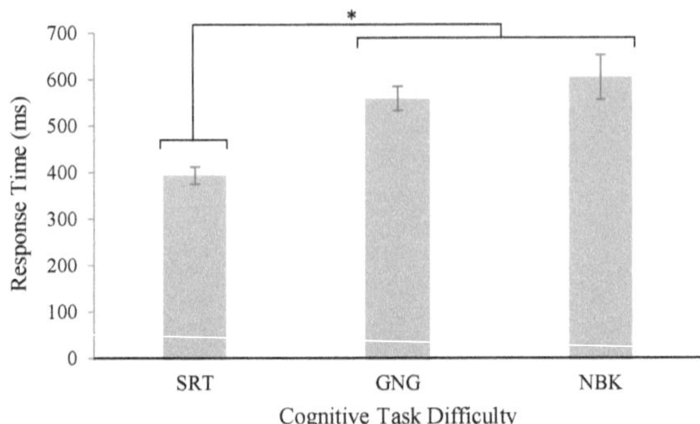

Figure 5. Mean response time (ms) changes between cognitive task difficulty levels (SRT), go/no-go (GNG) and n-back (NBK) SRT $F (2, 38) = 16.0$, $p < 0.001$, $\eta^2 = 0.456$. Response times in the GNG and NBK were significantly slower than the SRT ($p < 0.001$). (*) indicates significance $p < 0.001$. Error bars represent standard error of the mean.

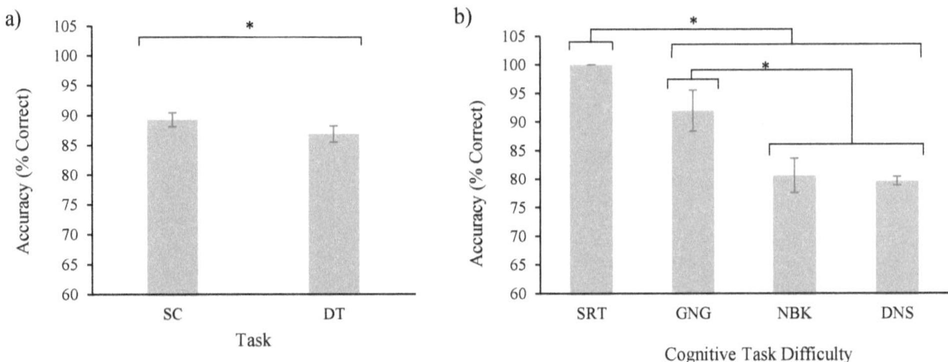

Figure 6. a) Mean accuracy decrease between single cognitive (SC) and dual-task (DT) $F (1, 19) = 5.7$, $p = 0.028$, $\eta^2 = 0.230$. SC was significantly more accurate than DT ($p < 0.001$). b) Mean decrease in accuracy (% correct) across cognitive task difficulty levels including simple reaction time (SRT), go/no-go (GNG), n-back (NBK) and double number sequence (DNS) $F (3, 57) = 16.2$, $p < 0.001$, $\eta^2 = 0.460$. Participants were significantly more accurate during the SRT than the GNG ($p = 0.038$), NBK ($p < 0.001$) and DNS ($p < 0.001$), and in the GNG compared to NBK ($p = 0.042$) and DNS ($p = 0.002$). (*) indicates significance $p < 0.05$. Error bars represent standard error of the mean.

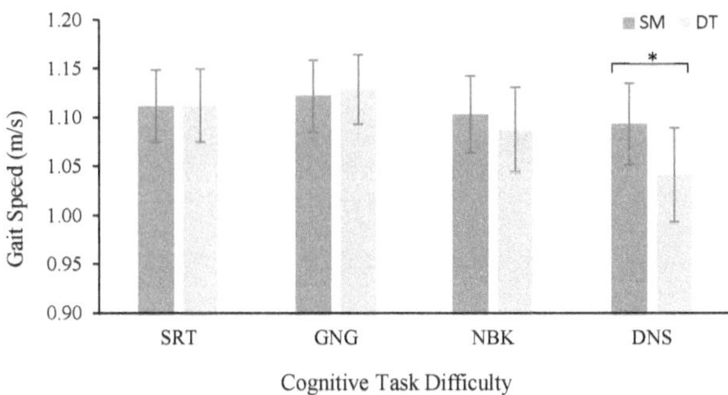

Figure 7. Mean gait speed changes (m/s) between single motor (SM) and dual-task (DT) blocks and across cognitive task difficulty levels $F(3, 57) = 2.2$, $p = 0.014$, $\eta^2 = 0.169$. Cognitive tasks include simple reaction time (SRT), go/no-go (GNG), n-back (NBK) and double number sequence (DNS). Mean gait speed was significantly slower between the DNS single- and dual-task ($p = 0.003$). (*) indicates significance $p < 0.05$. Error bars represent standard error of the mean.

CHAPTER FOUR: GENERAL DISCUSSION

The purpose of this study was to examine the changes in prefrontal cerebral oxygenation and behaviour in older adults during dual-task gait with increasing cognitive demands. More specifically, these changes were investigated as a function of cognitive task difficulty to determine whether different executive function domains are differentially affected by cognitive aging. Reports in the literature describe the relationship between brain activation and performance in terms of efficiency and compensation mechanisms (Cabeza et al., 2018; Causse et al., 2017; Reuter-Lorenz & Park, 2014). In other words, a cognitively demanding task that is processed efficiently may lead to enhanced performance. However, when cognitive demands exceed the available processing capacity, compensatory mechanisms, such as increased PFC activation, are initiated in an attempt to maintain performance. The dual-task paradigm can, therefore, be used to assess how efficiently older adults allocate their processing resources during two attention-demanding tasks (Pashler, 1994).

It was hypothesized that older adults would demonstrate increased brain activation between single- and dual-tasks and with increasing cognitive demand as outlined in compensation theories (Reuter-Lorenz & Park, 2014). Similarly, behavioural measures of performance including gait speed, response time and response accuracy were expected to decrease across single- and dual-tasks and with increasing cognitive task difficulty. Lastly, it was hypothesized that during the dual-tasks, increased brain activation would be correlated with decreased performance such that compensatory activation would be inefficient at maintaining performance with increasing cognitive demands.

4.1 Cerebral oxygenation changes across single- and dual-tasks

The neural findings from this study demonstrated a main effect of task such that cerebral oxygenation decreased from single- to dual-task. These findings partially support our initial hypothesis which was formulated according to neural compensation theories (Cabeza et al., 2018; Reuter-Lorenz & Park, 2014). The neuroimaging literature has demonstrated that compensatory patterns of neural activation include bilateral and more widespread brain activation in older compared to younger adults (Cabeza et al., 2002). Bilateral activation was demonstrated in our study in that there were no significant hemispheric differences in cerebral oxygenation between the left and right hemispheres during the different tasks and difficulty levels. As such, changes in brain activation were evaluated across the whole PFC. In terms of greater activation, our study did not find significant increases across cognitive task difficulty or between single- and dual-tasks. Rather, similar to Beurkens et al., (2014) there was a significant decrease in cerebral oxygenation between single- and dual-tasks.

Firstly, a decrease in cerebral oxygenation between single- and dual-tasks may represent an efficiency pattern of neural activation (Causse et al., 2017; Grady, 2012). Compared to walking alone, walking while executing a cognitive task is more challenging because of the competing demands of two simultaneous tasks. The increased attentional demands combined with the overlap of processing resources engage executive functions to a greater extent during dual-tasks. However, a decrease in cerebral oxygenation, as demonstrated in this study, may represent a more effective allocation of cognitive resources to allow executive processes to operate more efficiently. For example, Beurskens et al., (2014) demonstrated a decrease in prefrontal cerebral oxygenation between single- and dual-task walking. The authors of this study attributed their findings to the limited cognitive capacity of older adults (Reuter-Lorenz & Cappell, 2008). Therefore, when the older adults reached their maximum activation capacity, they may have redistributed their cognitive resources to other brain regions outside the PFC or used entirely different neural networks to process the task as efficiently as possible (Reuter-Lorenz & Cappell, 2008; Stern, 2009).

The reorganization of cognitive resources serves to optimize processing efficiency given the finite availability of prefrontal resources. Aging, in particular, can lead to reduced cerebral blood flow and may contribute to an overall reduced availability of prefrontal oxygenation (Agbangla et al., 2017; Bertsch et al., 2009). According to the theory of hypofrontality, certain tasks that elicit a greater demand for metabolic resources, such as acute bouts of exercise, which may induce a redistribution of metabolic resources away from the PFC (Dietrich & Audiffren, 2011). More specifically, acute bouts of exercise can be defined as a single session of exercise that can last between a few seconds to several hours (Dietrich & Audiffren, 2011). During this time, activities such as walking may require greater metabolic resources to optimize motor control as compared to executive processes that are involved in maintaining cognitive performance (Dietrich & Audiffren, 2011). Therefore, when individuals find themselves in complex situations that require processing beyond their available capacity, different brain regions may be prioritized based on their role in accomplishing a task. For this reason, an upregulation of metabolic resources towards motor regions and a downregulation of metabolic resources towards the PFC could be expected when walking is prioritized over cognitive performance. However, it is important to note that hypofrontality may be more beneficial for motor compared to cognitive performance (Dietrich & Audiffren, 2011). For example, cognitive tasks that involve many processing steps may be too complex to operate under minimal executive control. As such, a decrease in prefrontal resources may benefit motor performance or tasks that are more easily automated at the expense of cognitive performance.

Stuart et al., (2019) used fNIRS to examine the changes in cerebral oxygenation during dual-task walking in older adults. Findings from this study demonstrated that there were no significant differences in cerebral oxygenation in the PFC during single- versus dual-tasks. However, there was an increase in the premotor cortex, supplementary motor area and primary motor cortex during the dual-task. These regions, in particular, are involved in walking regulation such that the supplementary motor area is involved in the indirect locomotor pathway while the primary motor cortex is involved in the direct locomotor pathway (Harada et al., 2009; Herold et al., 2017; Lu et al., 2015; Miyai et al., 2001). Regions outside the PFC could not be measured within the scope of this study, however, we must acknowledge that other brain regions function simultaneously to facilitate motor performance during dual-task walking.

Lastly, in support of the cerebral oxygenation findings, there was a significant decrease in deoxygenated hemoglobin between single- and dual tasks. In fact, deoxygenated hemoglobin is strongly correlated with the fMRI BOLD response (Huppert et al., 2006). Changes in cerebral deoxygenation are less commonly reported in the fNIRS literature, however, it remains a reliable measure of neural activation. This is because HbR changes are typically harder to ascertain due to the low signal amplitude and often insignificant changes from the baseline measurements (Leff et al., 2011). However, the advantage of this measure is that it is less likely to be contaminated with physiological artifacts (Leff et al., 2011). As such, this study was able to capture cerebral deoxygenation changes that support and mirror the decrease observed in cerebral oxygenation.

4.2 Automatization of gait during dual-task walking

To further elaborate on the concepts of neural efficiency and compensation, the resulting motor performance that followed the decrease in cerebral oxygenation should be further examined. By first focusing on the changes in gait speed, the findings from this study demonstrated an interaction between single- and dual-tasks and cognitive task difficulty. More specifically, gait speed did not significantly differ across the first three levels of cognitive task difficulty, rather, it significantly decreased between single- and dual-tasks during the most difficult DNS working memory task. Although we expected gait speed to decrease between single- and dual-tasks and with increasing cognitive demand, gait speed maintenance across the first three levels of difficulty may represent an efficiency strategy in which gait performance was maintained despite decreased prefrontal brain activation. One model that supports these results is an automatic locomotor strategy, which involves the indirect locomotor control pathway (Herold et al., 2017). This pathway is advantageous in that it can function under minimal conscious attention (Bernstein, 1967; Poldrack, 2005). More importantly, it allows for gait speed maintenance under complex cognitive demands, such as dual-tasks, because it circumvents controlled processing in the PFC. There are certain features in this study that make walking more amenable to automation. For example, gait is a leant skill and older adults may prioritize motor over cognitive performance. However, the decrease in gait speed during the most difficult cognitive task may represent a shift from an automatic to an executive control strategy. More specifically, gait may require greater cognitive control when cognitive demands become too overwhelming for automatic control.

Firstly, walking may be easily automatized because it is a well-learnt skill (Schneider & Chein, 2003). Learnt tasks may not need to be processed in brain regions such as the PFC because they require minimal conscious attention to perform. Compared to the cognitive tasks used in this study, the motor task was more likely to be automatized because walking is a well-learnt skill that is used in everyday life. In addition, older adults, in particular, may achieve motor task automaticity following practice or training (Seidler et al., 2010; Wollesen & Voelcker-Rehage, 2014). For example, Wu et al., (2005) demonstrated that once older adults were extensively trained on a motor task, they could perform a cognitive task simultaneously. Since the motor task no longer required conscious attention, prefrontal resources could be reserved for the secondary task. As such, the older adults' prefrontal brain activity significantly decreased in comparison to the measurements acquired before training. Taken together with the gait speed findings from this study, since walking is well-learnt, brain activation can shift from prefrontal brain regions in the indirect locomotor pathway to the direct locomotor pathway to facilitate maintaining gait speed (Herold ct al., 2017). By automatically processing gait, executive resources can be reserved for processing more novel or unlearned tasks.

Following the dual-tasks, participants were asked a subjective question to determine how they allotted their attention during the dual-tasks. By rating how much they focused on the cognitive versus motor task out of a possible 100%, participants reported focusing <39% on the motor task. In other words, as the cognitive task demands increased, the older adults focused less on walking. The responses to this question may demonstrate an automatic locomotor control strategy in that the older adults reported paying less attention to walking even during the most cognitively demanding task. However, examining this from an alternate perspective, consciously focusing on the cognitive task may have equally facilitated automaticity. Studies have demonstrated that having an external focus offers advantages for postural control (Potvin-Desrochers et al., 2017; Wulf, 2013; Wulf et al., 2001). As suggested in the "constrained action hypothesis," focusing on the outcome of a movement (i.e., an external focus) causes less interference with automatic processes than focusing on the movement itself (i.e., an internal focus). Wulf et al., (2001) examined the constrained action hypothesis by instructing one group of participants to focus their balance on their feet while a second group focused on a marker placed a small distance away from their feet. The group that focused on the marker demonstrated better postural control than the first group. Similarly, diverting attention away from a postural task to a cognitive task even when cognitive demands are low may provide an external focus to improve motor performance (Huxhold et al., 2006). In our study, the cognitive tasks may have helped establish an alternate attentional focus. However, dual-tasks involve more complex processes that go beyond simply drawing attention away from the motor movement. In consequence, the participants were able to maintain their gait speed when certain dual-task processes also contributed to promoting automatic motor control.

In a review comparing healthy older adults to older adults with Parkinson's Disease (PD), older adults with PD adopted a "posture second" strategy by focusing attention on a cognitive task rather than balance stability (Bloem et al., 2006; Maidan et al., 2016). As a result, the participants with PD demonstrated impaired balance and greater gait variability during dual-tasks compared to healthy older adults. The findings from the present study demonstrated a "posture first" strategy in that gait speed was maintained across the first three levels of cognitive task difficulty. However, during the most demanding task, gait speed significantly declined between the single- and dual-tasks. According to the "posture first hypothesis," this may be the point in which the older adults felt unstable. Older adults have decreased muscle mass and may not be able to recover as easily from a fall as younger adults (Yogev-Seligmann et al., 2012). As such, the consequences of falling are far more severe than responding incorrectly to the cognitive task. Therefore, to accommodate for the change in cognitive demands, the older adults may have reduced their gait speed to ensure safe motor performance.

Interestingly, incrementally increasing cognitive task difficulty in our study may have demonstrated more subtle differences in gait speed maintenance that differentiate older and younger adults. Srygley et al., (2009) compared gait speed between the young and old age groups during an easy and difficult dual-task. Findings from this study demonstrated that older adults decreased their gait speed during the harder task while the younger adults maintained their speed throughout. This was partially observed in our study in that the older adults decreased their gait speed but only during the most cognitively demanding dual-task. However, employing various levels of cognitive task difficulty revealed that older adults may be able to maintain their gait speed when cognitive demands are relatively low. Only once task demands become too great will healthy older adults slow their gait speed, in line with the posture first hypothesis, to ensure walking stability.

4.3 Cognitive performance: response time and accuracy

Response time may be used to assess how older adults adapt their cognitive performance during attention-demanding tasks. Unlike gait speed, response time and accuracy were not maintained across single- and dual-tasks and task difficulty. More specifically, response times were significantly slower during the GNG and NBK tasks as compared to the SRT task. Since the DNS task is a non-verbal task, a response time was not measured. This supports our initial hypothesis such that greater task difficulty elicited worse performance in older adults. As expected, the SRT processing speed task resulted in the fastest response times since it required the least amount of processing steps (Potvin-Desrochers et al., 2017). During the SRT, participants were only required to respond quickly to each auditory stimulus. In comparison, the GNG response times may have been slower due to an extra processing step that involves inhibiting a response to a "no-go" stimulus before responding to the target stimulus. Older adults, in particular, are slower at inhibiting and distinguishing between relevant versus irrelevant stimuli, which may result in slower response times (Hasher & Zacks, 1988; Hsieh et al., 2016). Finally, the NBK working memory task required participants to maintain and continuously update information as new numbers were being announced. This added an additional level of complexity to the task which may take more time to process before producing a response. It should be noted, however, that participants were instructed to respond as quickly as possible to the SRT and GNG stimuli but were not told to do so during the NBK task. Although this may have contributed to slower response times during the NBK task, the GNG and NBK response times were not significantly different despite this difference in instructions. Similarly, given the longer interstimulus intervals (ISI) during the SRT and GNG tasks, the instructions to respond quickly may have been necessary to prevent long response times that are reflective of the large ISI and not the participants' capacity to respond quickly and accurately. With these considerations in mind, the response time findings suggest that more complex processing steps take longer for older adults to process.

These findings are in line with Vermeij et al., (2012) who demonstrated that older adults decreased their response time as cognitive task difficulty increased. In addition, similar to the neural findings from this study, the older adults recruited both hemispheres throughout all the difficulty levels, which may be a compensation mechanism to maintain performance. However, this was an unsuccessful attempt at compensation given that cognitive performance decreased with increasing task difficulty.

Furthermore, accuracy findings revealed that the single-task responses were more accurate than the dual-tasks. These findings are in line with our initial hypothesis in which competing attentional demands of two simultaneous tasks may lead to worse performance on one or both tasks. In this case, a performance decline was observed during the cognitive task such that older adults made more errors during the dual-tasks. Similar findings have been reported in the literature whereby older adults make more errors during complex cognitive-motor dual-tasks (Brustio et al., 2017; Srygley et al., 2009). Response accuracy also decreased with increasing cognitive task difficulty. More specifically, the SRT task was significantly more accurate than all the other difficulty levels and the GNG task was significantly more accurate than the NBK and DNS. However, the NBK and DNS tasks were not significantly different from one another. This may be because the NBK and DNS tasks involve elements of working memory and similar cognitive processes (Baddeley, 1986). The overall decline of accuracy with increasing task difficulty is in line with the literature which demonstrates that older adults make more errors as task demands increase (Fraser et al., 2016; Srygley et al., 2009; Vaportzis et al., 2013; Vermeij et al., 2012a). These findings support our difficulty manipulation such that cognitive performance decreased as task demands increased and differed across different executive function domains.

4.4 High and low performing older adults

Executive functioning may vary across different groups of older adults for reasons such as cognitive and physical status (Verghese & Xue, 2010). To account for this, studies evaluating this age group typically employ different neuropsychological tests and health questionnaires to determine the baseline characteristics of the older adults. Based on the participants' responses, some study protocols may choose to further categorize healthy older adults into high and low performing groups to better understand whether each group exhibits differences in cognitive and motor performance. The neuropsychological test scores obtained from this study demonstrated that the older adults possessed certain characteristics that pertain to high performing older adults. For example, one of the inclusion criteria for this study was a MoCA score ≥ 26 which is used as a cut-off for healthy cognition. In addition, compared to other studies, our participants were highly educated which is a factor that is often associated with more efficient processing speed in older adults (Clark et al., 2016). Similarly, our participants reported a low fear of falling compared to other studies that have associated fear of falling scores with greater PFC activation (Holtzer et al., 2019; Verghese et al., 2017).

The most prominent difference, however, between our study and others in the literature was the older adults' above average single- and dual-task gait speed (Smith et al., 2016). Verghese et al., (2017) define high functioning older adults as individuals who demonstrate clinically normal gait. Therefore, it can be argued that based on the single-task gait speed and the Short Physical Performance Battery score, our participants were physically healthy and could perform motorically at a clinically normal level. Furthermore, across all cognitive task difficulty levels, gait speed remained >1 m/s, which is a clinical marker of healthy gait speed in older adults (Hollman, McDade, et al., 2011; Kyrdalen et al., 2019; Verghese et al., 2017). In comparison, studies that have examined older adults who do not exhibit clinically normal gait have revealed that abnormal gait leads to increased prefrontal cerebral oxygenation and greater reliance on executively controlled processes (Harada et al., 2009; Maidan et al., 2016). Therefore, the gait speed findings from this study are in line with higher performing older adults who do not necessarily require extensive executive control while walking.

Furthermore, high and low performers may be distinguished using cognitive task performance. For example, Vermeij et al., (2014) divided older adults into high and low functioning groups based on their response accuracy during a working memory dual-task. In our study, the older adults were able to maintain accuracy levels $>80\%$ across all the levels of cognitive task difficulty. Therefore, these findings may be interpreted as a high level of performance throughout all the difficulty levels. In addition, it may account for studies that attribute a decrease in cerebral oxygenation to older adults disengaging or being disinterested in the dual-task because of the level of difficulty. Overall, the participants were able to function at a high level to complete all the tasks with a negligible number of blocks being removed from analyses.

4.5 Differences between dual-task study designs

Numerous studies have demonstrated that older adults exhibit greater cerebral oxygenation between single- and dual-tasks and across cognitive task difficulty (Pelicioni et al., 2019). However, an important distinction between those studies and our study is the use of an electronic walkway to examine gait speed (Hernandez et al., 2016; Holtzer et al., 2015; Mirelman et al., 2017; Verghese et al., 2017). While electronic walkways provide valuable information on gait performance, it may limit other analyses due to its restricted length. To overcome this limitation, our study evaluated over-ground walking along a 10 m walkway. This distance was designed to maximize straight line walking and is considerably longer than electronic walkways which typically range between 4-7 m. Differences in brain activation and automaticity may thus arise from the length of the walkway. For example, studies that have investigated brain activation during obstacle negation have demonstrated that the interruption of steady state walking leads to increased prefrontal activation (Hawkins et al., 2018; Holtzer et al., 2016; Maidan et al., 2016). Further, 10 m walkways provide a more reliable measure of gait speed than 5 m walkways (van Herk et al., 1998). More specifically, older adults, in particular, may require a walkway of at least eight meters for slower walkers and nine meters for faster walkers to accommodate for acceleration and deceleration phases (Macfarlane & Looney, 2008). The length of our walkway may have, therefore, contributed to our cognitive and motor performance findings which do not fully align with other dual-task findings that demonstrate increases in cerebral oxygenation and slower gait compared to single-tasks (Hernandez et al., 2016; Holtzer et al., 2015; Mirelman et al., 2017; Verghese et al., 2017).

The novelty of the cognitive tasks that were used in this study may have also contributed to our findings. The most commonly reported cognitive tasks in the literature are verbal fluency (Hawkins et al., 2018; Holtzer et al., 2015; Verghese et al., 2017) and counting backwards (Al-Yahya et al., 2016; Mirelman et al., 2017). A key feature of those tasks is that participants are given a starting cue (i.e., a number or letter) after which they can deliver responses at their own pace. This is important to consider as the dual-task paradigm assesses the allocation of cognitive resources during two attention-demanding tasks. Further, as outlined in the constrained action hypothesis, greater attention on an external focus, such as a cognitive task, may contribute to better postural control. Our study incorporated an unpredictable pattern of stimuli which prompted participants to respond or manipulate information throughout the block. This ensured that the participants were consistently engaged in both tasks and may have contributed to automatic locomotor control.

Our study was also designed to minimize any learning effects throughout the cognitive tasks. Firstly, the single- and dual-task blocks were presented in a counterbalanced order to minimize any learning effects. Similarly, the cognitive tasks were presented to each participant in a random order to ensure the were no difficulty effects based on which task was presented first. As mentioned previously, the more a task is learned, the more likely it is to be automatized. Since there were no significant differences between when the cognitive tasks were presented (e.g., SRT first versus last) and the order of each task (e.g., SC first versus last), it is unlikely that practicing the cognitive tasks led to improved learning as the experiment went on. In addition, there were no significant differences between testing days such that participants who completed the dynamic day first (walking dual-tasks) did not perform differently than those who completed the static day first (standing balance dual-tasks). More specifically, additional statistical analyses on the order of testing days, demonstrated that exposure to the cognitive tasks on Day 1 did not significantly alter performances on Day 2.

Lastly, we must account for differences within our own cognitive tasks. Each task was composed of a sequence of auditory stimuli that would prompt participants to either respond immediately or at the end of the block. However, compared to the SRT, GNG and NBK tasks, the DNS task was a non-verbal task in which participants withheld their responses until the end of the block. Therefore, we must consider the effect of speech on cerebral oxygenation which may be more present in the first three cognitive tasks compared to the DNS. Mirelman et al., (2014) examined the effect of verbalization on cerebral oxygenation during dual-task walking. The authors compared four conditions: normal walking, walking while counting forward, walking while serially subtracting seven and standing while serially subtracting seven. Findings from this study demonstrated that cerebral oxygenation increased with increasing task difficulty. However, standing and serially subtracting seven resulted in a negative change in cerebral oxygenation. Therefore, the significant difference between each verbalization condition was attributed to task difficulty rather than speech. This is supported by Holtzer et al., (2015) who demonstrated that although the left hemisphere is used to regulate speech, there was an equal increase of cerebral oxygenation in both the left and right hemispheres. Our findings demonstrated similar activation in both hemispheres and there were no significant differences in cerebral oxygenation between the verbal and non-verbal tasks. Therefore, it is unlikely that speech affected cerebral oxygenation in our study.

Nonetheless, our cognitive tasks minimized speech by having participants respond with a number between 0-9 or with the word "top." Both types of responses are made up of short, one or two syllable words to minimize the amount of speaking during the task. In addition, participants were not speaking as consistently throughout the block as what would be expected in a verbal fluency task. However, speech may still appear as a noise artifact in the cerebral oxygenation signal (Scholkmann et al., 2014; Tachtsidis & Scholkmann, 2016). To account for this, male and female speech frequencies fall within the region >120 Hz, which would be filtered out during the pre-processing stages using a bandpass filter (Simpson, 2009).

CHAPTER FIVE: CONCLUSION

5.1 Summary of findings

Findings from this study helped reveal the PFC's role in different executive function tasks during dual-task gait. Neural and behavioural findings demonstrated that simultaneously performing two attention-demanding tasks resulted in changes in brain activation and performance in older adults. From a neural perspective, brain activation decreased between single- and dual-tasks which may be interpreted as a more efficient processing strategy. Taken together with gait speed findings, this may be an example of automatic locomotor control such that gait speed was maintained up until the most difficult cognitive task. However, upon reaching this level of difficulty, gait speed significantly decreased which may represent a shift from automatic to executive control. Furthermore, cognitive performance measures including response time and accuracy decreased significantly across task difficulty. More specifically, response times were slower based on the level of processing complexity that was required to execute the cognitive task. Accuracy also decreased between single- and dual-tasks and across task difficulty. These findings support our difficulty manipulation and demonstrate that older adults were not able to maintain cognitive performance during the dual-tasks. There were no significant correlations between dual-task PFC activation and performance. However, based on the neural and behavioural findings, the older adults were better able to manage their motor compared to cognitive performance during the dual-tasks.

5.2 Clinical implications and future directions

The results of this study helped contribute to our understanding of automatic and executively controlled processes in healthy older adults. When cognitive demands are low, older adults may be able to maintain motor performance at the expense of cognitive performance by automatizing gait. Furthermore, the efficiency of certain cognitive processes can be inferred using evidence from behavioural measures such as response time, accuracy and gait speed. For example, when older adults demonstrate increased brain activation but decreased cognitive or motor performance, this may be a sign of inefficiency and deficits in executive functioning.

In a clinical setting, automaticity is commonly used to assess walking abilities (Middleton et al., 2015). Therefore, when there is a loss of automaticity, older adults may demonstrate slower gait speed. Gait speed changes can be further examined using a dual-task intervention to assess whether or not older adults can effectively allocate their attention to maintain motor performance (Verghese et al., 2017). By employing different levels of cognitive task difficulty, this may reveal whether performance decreases across various executive function domains or whether an intervention can be tailored to improve a certain performance outcome (Brustio et al., 2018; Wollesen et al., 2017). However, dual-tasking may not be recommended for older adults who do not show improvements following training or who continue to display performance deficits that are below the expectation for their age.

Few studies have been able to simultaneously examine the PFC and the other motor regions that are involved in the direct locomotor pathway. Those studies largely focused on younger adults (Koenraadt et al., 2014; Lu et al., 2015; Suzuki et al., 2008) or involved treadmill walking rather than over ground walking (Harada et al., 2009; Stuart et al., 2019). With the recent advancements in fNIRS, including greater portability and more extensive coverage of the brain, future studies should consider examining multiple brain regions simultaneously. More specifically, examining brain activation and performance in older adults across different task difficulties may provide greater insights into the balance between automatic and controlled processes. This may lead to more novel approaches to examine executive functions that can be used in more realistic situations outside the laboratory.

5.3 Interdisciplinarity of study

The central focus of this study involved the interplay between cognition and walking performance in older adults. As such, the present study incorporated cognitive science and human kinetics perspectives to better understand the cognitive-motor pathways involved in dual-task walking. Additionally, this study was part of a broader examination of static balance in older adults that incorporated a large interdisciplinary team for data collection and an engineer to facilitate the fNIRS signal processing. This created an environment for rich interdisciplinary exchange and learning opportunities among professors and students. The knowledge gained from neural and behavioural performance measures will help to inform both human kinetics and cognitive science fields.

REFERENCES

Agbangla, N., Audiffren, M., & Albinet, C. (2017). Use of near-infrared spectroscopy in the investigation of brain activation during cognitive aging: A systematic review of an emerging area of research. *Ageing Research Reviews, 38,* 52–66. https://doi.org/10.1016/j.arr.2017.07.003

Albinet, C., Boucard, G., Bouquet, C., & Audiffren, M. (2012). Processing speed and executive functions in cognitive aging: How to disentangle their mutual relationship? *Brain and Cognition, 79*(1), 1–11. https://doi.org/10.1016/j.bandc.2012.02.001

Al-Yahya, E., Dawes, H., Smith, L., Dennis, A., Howells, K., & Cockburn, J. (2011). Cognitive motor interference while walking: A systematic review and meta-analysis. *Neuroscience & Biobehavioral Reviews, 35*(3), 715–728. https://doi.org/10.1016/j.neubiorev.2010.08.008

Al-Yahya, E., Johansen-Berg, H., Kischka, U., Zarei, M., Cockburn, J., & Dawes, H. (2016). Prefrontal Cortex Activation While Walking Under Dual-Task Conditions in Stroke: A Multimodal Imaging Study. *Neurorehabilitation and Neural Repair, 30*(6), 591–599. https://doi.org/10.1177/1545968315613864

Baddeley, A. (1986). *Working memory.* Oxford: Oxford University Press.

Beck, E., Intzandt, B., & Almeida, Q. (2018). Can dual task walking improve in Parkinson's disease after external focus of attention exercise? A single blind randomized controlled trial. *Neurorehabilitation and Neural Repair, 32*(1), 18–33. https://doi.org/10.1177/1545968317746782

Bernstein, N. (1967). *The co-ordination and regulation of movements.* Oxford, New York, Pergamon Press.

Bertsch, K., Hagemann, D., Hermes, M., Walter, C., Khan, R., & Naumann, E. (2009). Resting cerebral blood flow, attention, and aging. *Brain Research*, *1267*, 77–88. https://doi.org/10.1016/j.brainres.2009.02.053

Beurskens, R., & Bock, O. (2012). Age-related deficits of dual-task walking: A review. *Neural Plasticity*, *2012*. https://doi.org/10.1155/2012/131608

Beurskens, R., Helmich, I., Rein, R., & Bock, O. (2014). Age-related changes in prefrontal activity during walking in dual-task situations: A fNIRS study. *International Journal of Psychophysiology*, *92*(3), 122–128. https://doi.org/10.1016/j.ijpsycho.2014.03.005

Bloem, B., Grimbergen, Y., van Dijk, J., & Munneke, M. (2006). The "posture second" strategy: A review of wrong priorities in Parkinson's disease. *Journal of the Neurological Sciences*, *248*(1–2), 196–204. https://doi.org/10.1016/j.jns.2006.05.010

Bohannon, R., Andrews, A., & Thomas, M. (1996). Walking speed: Reference values and correlates for older adults. *Journal of Orthopaedic & Sports Physical Therapy*, *24*(2), 86–90. https://doi.org/10.2519/jospt.1996.24.2.86

Broadbent, D. (1958). *Perception and communication*. Pergamon Press.

Brustio, P., Magistro, D., Zecca, M., Rabaglietti, E., & Liubicich, M. (2017). Age-related decrements in dual-task performance: Comparison of different mobility and cognitive tasks. A cross sectional study. *PLOS ONE*, *12*(7), e0181698. https://doi.org/10.1371/journal.pone.0181698

Brustio, P., Rabaglietti, E., Formica, S., & Liubicich, M. (2018). Dual-task training in older adults: The effect of additional motor tasks on mobility performance. *Archives of Gerontology and Geriatrics*, *75*, 119–124. https://doi.org/10.1016/j.archger.2017.12.003

Bruya, B. (2010). *Effortless attention: A new perspective in the cognitive science of attention and action*. https://books-scholarsportal-info.proxy.bib.uottawa.ca/uri/ebooks/ebooks3/oso/2013-08-08/1/upso-9780262013840-Bruya

Bürki, C., Bridenbaugh, S., Reinhardt, J., Stippich, C., Kressig, R., & Blatow, M. (2017). Imaging gait analysis: An fMRI dual task study. *Brain and Behavior*, *7*(8), e00724. https://doi.org/10.1002/brb3.724

Cabeza, R., Albert, M., Belleville, S., Craik, F., Duarte, A., Grady, C., Lindenberger, U., Nyberg, L., Park, D., Reuter-Lorenz, P., Rugg, M., Steffener, J., & Rajah, M. (2018). Maintenance, reserve and compensation: The cognitive neuroscience of healthy ageing. *Nature Reviews Neuroscience*. https://doi.org/10.1038/s41583-018-0068-2

Cabeza, R., Anderson, N., Locantore, J., & McIntosh, A. (2002). Aging gracefully: compensatory brain activity in high-performing older adults. *NeuroImage*, *17*(3), 1394–1402. https://doi.org/10.1006/nimg.2002.1280

Cabeza, R., & Dennis, N. (2013). Frontal lobes and aging. In D. T. Stuss & R. T. Knight (Eds.), *Principles of Frontal Lobe Function* (pp. 628–652). Oxford University Press. https://doi.org/10.1093/med/9780199837755.003.0044

Causse, M., Chua, Z., Peysakhovich, V., Del Campo, N., & Matton, N. (2017). Mental workload and neural efficiency quantified in the prefrontal cortex using fNIRS. *Scientific Reports*, *7*(1). https://doi.org/10.1038/s41598-017-05378-x

Clark, D. (2015). Automaticity of walking: Functional significance, mechanisms, measurement and rehabilitation strategies. *Frontiers in Human Neuroscience*, *9*. https://doi.org/10.3389/fnhum.2015.00246

Clark, D., Xu, H., Unverzagt, F., & Hendrie, H. (2016). Does targeted cognitive training reduce educational disparities in cognitive function among cognitively normal older adults? *International Journal of Geriatric Psychiatry*, *31*(7), 809–817. https://doi.org/10.1002/gps.4395

Deary, I., & Der, G. (2005). Reaction time, age, and cognitive ability: Longitudinal findings from age 16 to 63 years in representative population samples. *Aging, Neuropsychology, and Cognition*, *12*(2), 187–215. https://doi.org/10.1080/13825580590969235

Delbaere, K., Close, J., Mikolaizak, A., Sachdev, P., Brodaty, H., & Lord, S. (2010). The falls efficacy scale international (FES-I). A comprehensive longitudinal validation study. *Age and Ageing*, *39*(2), 210–216. https://doi.org/10.1093/ageing/afp225

Diamond, A. (2013). Executive functions. *Annual Review of Psychology*, *64*, 135–168. https://doi.org/10.1146/annurev-psych-113011-143750

Dietrich, A. (2003). Functional neuroanatomy of altered states of consciousness: The transient hypofrontality hypothesis. *Consciousness and Cognition*, *12*(2), 231–256. https://doi.org/10.1016/S1053-8100(02)00046-6

Dietrich, A., & Audiffren, M. (2011). The reticular-activating hypofrontality (RAH) model of acute exercise. *Neuroscience & Biobehavioral Reviews*, *35*(6), 1305–1325. https://doi.org/10.1016/j.neubiorev.2011.02.001

Dupuy, O., Gauthier, C., Fraser, S., Desjardins-Crèpeau, L., Desjardins, M., Mekary, S., Lesage, F., Hoge, R., Pouliot, P., & Bherer, L. (2015). Higher levels of cardiovascular fitness are associated with better executive function and prefrontal oxygenation in younger and older women. *Frontiers in Human Neuroscience*, *9*. https://doi.org/10.3389/fnhum.2015.00066

Düzel, E., Schütze, H., Yonelinas, A., & Heinze, H. (2010). Functional phenotyping of successful aging in long-term memory: Preserved performance in the absence of neural compensation. *Hippocampus*, n/a-n/a. https://doi.org/10.1002/hipo.20834

Forstmann, B., Tittgemeyer, M., Wagenmakers, E., Derrfuss, J., Imperati, D., & Brown, S. (2011). The speed-accuracy tradeoff in the elderly brain: a structural model-based approach. *Journal of Neuroscience, 31*(47), 17242–17249. https://doi.org/10.1523/JNEUROSCI.0309-11.2011

Fraser, S., Dupuy, O., Pouliot, P., Lesage, F., & Bherer, L. (2016). Comparable Cerebral oxygenation patterns in younger and older adults during dual-task walking with increasing load. *Frontiers in Aging Neuroscience, 8*. https://doi.org/10.3389/fnagi.2016.00240

Grady, C. (2012). The cognitive neuroscience of ageing. *Nature Reviews Neuroscience, 13*(7), 491–505. https://doi.org/10.1038/nrn3256

Grady, C. (2000). Functional brain imaging and age-related changes in cognition. *Biological Psychology, 54*(1), 259–281. https://doi.org/10.1016/S0301-0511(00)00059-4

Guralnik, J., Simonsick, E., Ferrucci, L., Glynn, R., Berkman, L., Blazer, D., Scherr, P., & Wallace, R. (1994). A Short physical performance battery assessing lower extremity function: association with self-reported disability and prediction of mortality and nursing home admission. *Journal of Gerontology, 49*(2), M85–M94. https://doi.org/10.1093/geronj/49.2.M85

Hamacher, D., Herold, F., Wiegel, P., Hamacher, D., & Schega, L. (2015). Brain activity during walking: A systematic review. *Neuroscience & Biobehavioral Reviews, 57*, 310–327. https://doi.org/10.1016/j.neubiorev.2015.08.002

Harada, T., Miyai, I., Suzuki, M., & Kubota, K. (2009). Gait capacity affects cortical activation patterns related to speed control in the elderly. *Experimental Brain Research, 193*(3), 445–454. https://doi.org/10.1007/s00221-008-1643-y

Hasher, L., & Zacks, R. (1988). Working memory, comprehension, and aging: A review and a new view. In G. H. Bower (Ed.), *Psychology of Learning and Motivation* (Vol. 22, pp. 193–225). Academic Press. https://doi.org/10.1016/S0079-7421(08)60041-9

Hausdorff, J., Schweiger, A., Herman, T., Yogev-Seligmann, G., & Giladi, N. (2008). Dual task decrements in gait among healthy older adults: contributing factors. *The Journals of Gerontology. Series A, Biological Sciences and Medical Sciences, 63*(12), 1335–1343.

Hawkins, K., Fox, E., Daly, J., Rose, D., Christou, E., McGuirk, T., Otzel, D., Butera, K., Chatterjee, S., & Clark, D. (2018). Prefrontal over-activation during walking in people with mobility deficits: Interpretation and functional implications. *Human Movement Science, 59*, 46–55. https://doi.org/10.1016/j.humov.2018.03.010

Hernandez, M., Holtzer, R., Chaparro, G., Jean, K., Balto, J., Sandroff, B., Izzetoglu, M., & Motl, R. (2016). Brain activation changes during locomotion in middle-aged to older adults with multiple sclerosis. *Journal of the Neurological Sciences, 370*, 277–283. https://doi.org/10.1016/j.jns.2016.10.002

Herold, F., Wiegel, P., Scholkmann, F., Thiers, A., Hamacher, D., & Schega, L. (2017). Functional near-infrared spectroscopy in movement science: A systematic review on cortical activity in postural and walking tasks. *Neurophotonics, 4*(4). https://doi.org/10.1117/1.NPh.4.4.041403

Herwig, U., Satrapi, P., & Schönfeldt-Lecuona, C. (2003). Using the international 10-20 eeg system for positioning of transcranial magnetic stimulation. *Brain Topography*, *16*(2), 95–99. https://doi.org/10.1023/B:BRAT.0000006333.93597.9d

Hollman, J., McDade, E., & Petersen, R. (2011). Normative spatiotemporal gait parameters in older adults. *Gait & Posture*, *34*(1), 111–118. https://doi.org/10.1016/j.gaitpost.2011.03.024

Hollman, J., Youdas, J., & Lanzino, D. (2011). Gender differences in dual task gait performance in older adults. *American Journal of Men's Health*, *5*(1), 11–17. https://doi.org/10.1177/1557988309357232

Holtzer, R., Kraut, R., Izzetoglu, M., & Ye, K. (2019). The effect of fear of falling on prefrontal cortex activation and efficiency during walking in older adults. *GeroScience*, *41*(1), 89–100. https://doi.org/10.1007/s11357-019-00056-4

Holtzer, R., Mahoney, J., Izzetoglu, M., Izzetoglu, K., Onaral, B., & Verghese, J. (2011). FNIRS study of walking and walking while talking in young and old individuals. *The Journals of Gerontology Series A: Biological Sciences and Medical Sciences*, *66A*(8), 879–887. https://doi.org/10.1093/gerona/glr068

Holtzer, R., Mahoney, J., Izzetoglu, M., Wang, C., England, S., & Verghese, J. (2015). Online fronto-cortical control of simple and attention-demanding locomotion in humans. *NeuroImage*, *112*, 152–159. https://doi.org/10.1016/j.neuroimage.2015.03.002

Holtzer, R., Verghese, J., Allali, G., Izzetoglu, M., Wang, C., & Mahoney, J. (2016). Neurological gait abnormalities moderate the functional brain signature of the posture first hypothesis. *Brain Topography*, *29*(2), 334–343. https://doi.org/10.1007/s10548-015-0465-z

Holtzer, R., Verghese, J., Xue, X., & Lipton, R. (2006). Cognitive processes related to gait velocity: Results from the Einstein aging study. *Neuropsychology, 20*(2), 215–223. https://doi.org/10.1037/0894-4105.20.2.215

Hsieh, S., Wu, M., & Tang, C. (2016). Adaptive strategies for the elderly in inhibiting irrelevant and conflict no-go trials while performing the go/no-go task. *Frontiers in Aging Neuroscience, 7*. https://doi.org/10.3389/fnagi.2015.00243

Huppert, T., Hoge, R., Diamond, S., Franceschini, M., & Boas, D. (2006). A temporal comparison of BOLD, ASL, and NIRS hemodynamic responses to motor stimuli in adult humans. *NeuroImage, 29*(2), 368–382. https://doi.org/10.1016/j.neuroimage.2005.08.065

Huxhold, O., Li, S., Schmiedek, F., & Lindenberger, U. (2006). Dual-tasking postural control: Aging and the effects of cognitive demand in conjunction with focus of attention. *Brain Research Bulletin, 69*(3), 294–305. https://doi.org/10.1016/j.brainresbull.2006.01.002

Jöbsis, F. (1977). Noninvasive, infrared monitoring of cerebral and myocardial oxygen sufficiency and circulatory parameters. *Science, 198*(4323), 1264–1267. JSTOR.

Kahya, M., Moon, S., Ranchet, M., Vukas, R., Lyons, K., Pahwa, R., Akinwuntan, A., & Devos, H. (2019). Brain activity during dual task gait and balance in aging and age-related neurodegenerative conditions: A systematic review. *Experimental Gerontology*, 110756. https://doi.org/10.1016/j.exger.2019.110756

Kal, E., van der Kamp, J., Houdijk, H., Groet, E., van Bennekom, C., & Scherder, E. (2015). Stay focused! The effects of internal and external focus of attention on movement automaticity in patients with stroke. *PLOS ONE, 10*(8), e0136917. https://doi.org/10.1371/journal.pone.0136917

Kocsis, L., Herman, P., & Eke, A. (2006). The modified Beer–Lambert law revisited. *Physics in Medicine and Biology*, *51*(5), N91–N98. https://doi.org/10.1088/0031-9155/51/5/N02

Koenraadt, K., Roelofsen, E., Duysens, J., & Keijsers, N. (2014). Cortical control of normal gait and precision stepping: An fNIRS study. *NeuroImage*, *85*, 415–422. https://doi.org/10.1016/j.neuroimage.2013.04.070

Kyrdalen, I., Thingstad, P., Sandvik, L., & Ormstad, H. (2019). Associations between gait speed and well-known fall risk factors among community-dwelling older adults. *Physiotherapy Research International*, *24*(1), e1743. https://doi.org/10.1002/pri.1743

Lajoie, Y., Teasdale, N., Bard, C., & Fleury, M. (1996). Upright standing and gait: Are there changes in attentional requirements related to normal aging? *Experimental Aging Research*, *22*(2), 185–198. https://doi.org/10.1080/03610739608254006

Leff, D., Orihuela-Espina, F., Elwell, C., Athanasiou, T., Delpy, D., Darzi, A., & Yang, G. (2011). Assessment of the cerebral cortex during motor task behaviours in adults: A systematic review of functional near infrared spectroscopy (fNIRS) studies. *NeuroImage*, *54*(4), 2922–2936. https://doi.org/10.1016/j.neuroimage.2010.10.058

Lövdén, M., Schaefer, S., Pohlmeyer, A., & Lindenberger, U. (2008). Walking variability and working-memory load in aging: A dual-process account relating cognitive control to motor control performance. *The Journals of Gerontology: Series B*, *63*(3), P121–P128. https://doi.org/10.1093/geronb/63.3.P121

Lu, C., Liu, Y., Yang, Y., Wu, Y., & Wang, R. (2015). Maintaining gait performance by cortical activation during dual-task interference: A functional near-infrared spectroscopy study. *PLoS ONE*, *10*(6). https://doi.org/10.1371/journal.pone.0129390

Lundin-Olsson, L., Nyberg, L., Gustafson, Y., Himbert, D., Seknadji, P., Karila-Cohen, D., Juliard, J., & Steg, P. (1997). *"Stops walking when talking" as a predictor of falls in elderly people. 349*, 1.

Macfarlane, P., & Looney, M. (2008). Walkway length determination for steady state walking in young and older adults. *Research Quarterly for Exercise and Sport, 79*(2), 261–267.

Maidan, I., Nieuwhof, F., Bernad-Elazari, H., Reelick, M., Bloem, B., Giladi, N., Deutsch, J., Hausdorff, J., Claassen, J., & Mirelman, A. (2016). The Role of the frontal lobe in complex walking among patients with parkinson's disease and healthy older adults: An fNIRS study. *Neurorehabilitation and Neural Repair, 30*(10), 963–971. https://doi.org/10.1177/1545968316650426

Middleton, A., Fritz, S., & Lusardi, M. (2015). Walking speed: The functional vital sign. *Journal of Aging and Physical Activity, 23*(2), 314–322. https://doi.org/10.1123/japa.2013-0236

Mirelman, A., Maidan, I., Bernad-Elazari, H., Shustack, S., Giladi, N., & Hausdorff, J. (2017). Effects of aging on prefrontal brain activation during challenging walking conditions. *Brain and Cognition, 115*, 41–46. https://doi.org/10.1016/j.bandc.2017.04.002

Miyai, I., Tanabe, H., Sase, I., Eda, H., Oda, I., Konishi, I., Tsunazawa, Y., Suzuki, T., Yanagida, T., & Kubota, K. (2001). Cortical mapping of gait in humans: a near-infrared spectroscopic topography study. *NeuroImage, 14*(5), 1186–1192. https://doi.org/10.1006/nimg.2001.0905

Miyake, A., Friedman, N., Emerson, M., Witzki, A., Howerter, A., & Wager, T. (2000). The unity and diversity of executive functions and their contributions to complex "frontal lobe" tasks: A latent variable analysis. *Cognitive Psychology, 41*(1), 49–100. https://doi.org/10.1006/cogp.1999.0734

Nasreddine, Z., Phillips, N., Bédirian, V., Charbonneau, S., Whitehead, V., Collin, I., Cummings, J., & Chertkow, H. (2005). The Montreal cognitive assessment, MoCA: A brief screening tool for mild cognitive impairment. *Journal of the American Geriatrics Society, 53*(4), 695–699. https://doi.org/10.1111/j.1532-5415.2005.53221.x

Oldfield, R. (1971). The assessment and analysis of handedness: The Edinburgh inventory. *Neuropsychologia, 9*(1), 97–113. https://doi.org/10.1016/0028-3932(71)90067-4

Pashler, H. (1994). *Dual-task interference in simple tasks: data and theory. 116*(2), 25. https://doi.org/10.1037/0033-2909.116.2.220

Patel, P., Lamar, M., & Bhatt, T. (2014). Effect of type of cognitive task and walking speed on cognitive-motor interference during dual-task walking. *Neuroscience, 260*, 140–148. https://doi.org/10.1016/j.neuroscience.2013.12.016

Pelicioni, P., Tijsma, M., Lord, S., & Menant, J. (2019). Prefrontal cortical activation measured by fNIRS during walking: Effects of age, disease and secondary task. *PeerJ, 7*, e6833. https://doi.org/10.7717/peerj.6833

Pinti, P., Tachtsidis, I., Hamilton, A., Hirsch, J., Aichelburg, C., Gilbert, S., & Burgess, P. (2018). The present and future use of functional near-infrared spectroscopy (fNIRS) for cognitive neuroscience. *Annals of the New York Academy of Sciences, 0*(0), 1–25. https://doi.org/10.1111/nyas.13948

Poldrack, R. (2005). The neural correlates of motor skill automaticity. *Journal of Neuroscience, 25*(22), 5356–5364. https://doi.org/10.1523/JNEUROSCI.3880-04.2005

Potvin-Desrochers, A., Richer, N., & Lajoie, Y. (2017). Cognitive tasks promote automatization of postural control in young and older adults. *Gait & Posture, 57*, 40–45. https://doi.org/10.1016/j.gaitpost.2017.05.019

Quaresima, V., & Ferrari, M. (2019). A mini-review on functional near-infrared spectroscopy (fNIRS): Where do we stand, and where should we go? *Photonics*, *6*(3), 87. https://doi.org/10.3390/photonics6030087

Rabbitt, P. (1979). How old and young subjects monitor and control responses for accuracy and speed. *British Journal of Psychology*, *70*(2), 305–311. https://doi.org/10.1111/j.2044-8295.1979.tb01687.x

Raz, N., & Rodrigue, K. (2006). Differential aging of the brain: Patterns, cognitive correlates and modifiers. *Neuroscience and Biobehavioral Reviews*, *30*(6), 730–748. https://doi.org/10.1016/j.neubiorev.2006.07.001

Reuter-Lorenz, P., & Cappell, K. (2008). Neurocognitive Aging and the Compensation Hypothesis. *Current Directions in Psychological Science*, *17*(3), 177–182. https://doi.org/10.1111/j.1467-8721.2008.00570.x

Reuter-Lorenz, P., & Park, D. (2014). How does it STAC up? Revisiting the scaffolding theory of aging and cognition. *Neuropsychology Review*, *24*(3), 355–370. https://doi.org/10.1007/s11065-014-9270-9

Richer, N., Saunders, D., Polskaia, N., & Lajoie, Y. (2017). The effects of attentional focus and cognitive tasks on postural sway may be the result of automaticity. *Gait & Posture*, *54*, 45–49. https://doi.org/10.1016/j.gaitpost.2017.02.022

Rosso, A., Cenciarini, M., Sparto, P., Loughlin, P., Furman, J., & Huppert, T. (2017). Neuroimaging of an attention demanding dual-task during dynamic postural control. *Gait & Posture*, *57*, 193–198. https://doi.org/10.1016/j.gaitpost.2017.06.013

Sala, S., Baddeley, A., Papagno, C., & Spinnler, H. (1995). Dual-task paradigm: A means to examine the central executive. *Annals of the New York Academy of Sciences, 769*(1 Structure and), 161–172. https://doi.org/10.1111/j.1749-6632.1995.tb38137.x

Salthouse, T. A. (1996). *The processing-speed theory of adult age differences in cognition. 103*(3), 403–428.

Schneider, W., & Chein, J. (2003). Controlled & automatic processing: Behavior, theory, and biological mechanisms. *Cognitive Science, 27*(3), 525–559. https://doi.org/10.1207/s15516709cog2703_8

Schneider, W., & Shiffrin, R. (1977). *Controlled and automatic human information processing: I. detection, search, and attention. 66.*

Scholkmann, F., Kleiser, S., Metz, A., Zimmermann, R., Mata Pavia, J., Wolf, U., & Wolf, M. (2014). A review on continuous wave functional near-infrared spectroscopy and imaging instrumentation and methodology. *NeuroImage, 85,* 6–27. https://doi.org/10.1016/j.neuroimage.2013.05.004

Scholkmann, F., & Wolf, M. (2013). General equation for the differential pathlength factor of the frontal human head depending on wavelength and age. *Journal of Biomedical Optics, 18*(10), 105004. https://doi.org/10.1117/1.JBO.18.10.105004

Seidler, R., Bernard, J., Burutolu, T., Fling, B., Gordon, M., Gwin, J., Kwak, Y., & Lipps, D. (2010). Motor control and aging: Links to age-related brain structural, functional, and biochemical effects. *Neuroscience & Biobehavioral Reviews, 34*(5), 721–733. https://doi.org/10.1016/j.neubiorev.2009.10.005

Shumway-Cook, A., Woollacott, M., Kerns, K., & Baldwin, M. (1997). The effects of two types of cognitive tasks on postural stability in older adults with and without a history of falls.

The Journals of Gerontology Series A: Biological Sciences and Medical Sciences, *52A*(4), M232–M240. https://doi.org/10.1093/gerona/52A.4.M232

Simpson, A. (2009). Phonetic differences between male and female speech. *Language and Linguistics Compass*, *3*(2), 621–640. https://doi.org/10.1111/j.1749-818X.2009.00125.x

Smith, E., Cusack, T., & Blake, C. (2016). The effect of a dual task on gait speed in community dwelling older adults: A systematic review and meta-analysis. *Gait & Posture*, *44*, 250–258. https://doi.org/10.1016/j.gaitpost.2015.12.017

Smith, E., Cusack, T., Cunningham, C., & Blake, C. (2017). The influence of a cognitive dual task on the gait parameters of healthy older adults: A systematic review and meta-analysis. *Journal of Aging and Physical Activity*, *25*(4), 671–686. https://doi.org/10.1123/japa.2016-0265

Sorond, F., Kiely, D., Galica, A., Moscufo, N., Serrador, J., Iloputaife, I., Egorova, S., Dell'Oglio, E., Meier, D. S., Newton, E., Milberg, W., Guttmann, C., & Lipsitz, L. (2011). Neurovascular coupling is impaired in slow walkers: The MOBILIZE Boston Study. *Annals of Neurology*, *70*(2), 213–220. https://doi.org/10.1002/ana.22433

Srygley, J., Mirelman, A., Herman, T., Giladi, N., & Hausdorff, J. (2009). When does walking alter thinking? Age and task associated findings. *Brain Research*, *1253*, 92–99. https://doi.org/10.1016/j.brainres.2008.11.067

St-Amant, G., Rahman, T., Polskaia, N., Fraser, S., & Lajoie, Y. (2020). Unveilling the cerebral and sensory contributions to automatic postural control during dual-task standing. *Human Movement Science*, *70*, 102587. https://doi.org/10.1016/j.humov.2020.102587

Stern, Y. (2009). Cognitive reserve. *Neuropsychologia*, *47*(10), 2015–2028. https://doi.org/10.1016/j.neuropsychologia.2009.03.004

Strauss, P., Sherman, N. & Spreen, O. (2006). *A compendium of neuropsychological tests: administration, norms, and commentary.* Oxford University Press.

Stroop, J. (1935). Studies of interference in serial verbal reactions. *Journal of Experimental Psychology, 18*(6), 643–662.

Stuart, S., Alcock, L., Rochester, L., Vitorio, R., & Pantall, A. (2019). Monitoring multiple cortical regions during walking in young and older adults: Dual-task response and comparison challenges. *International Journal of Psychophysiology, 135*, 63–72. https://doi.org/10.1016/j.ijpsycho.2018.11.006

Suzuki, M., Miyai, I., Ono, T., & Kubota, K. (2008). Activities in the frontal cortex and gait performance are modulated by preparation. An fNIRS study. *NeuroImage, 39*(2), 600–607. https://doi.org/10.1016/j.neuroimage.2007.08.044

Tachtsidis, I., & Scholkmann, F. (2016). False positives and false negatives in functional near-infrared spectroscopy: Issues, challenges, and the way forward. *Neurophotonics, 3*(3), 031405. https://doi.org/10.1117/1.NPh.3.3.031405

Tang, Z., & Wakayama, S. (2011). Age-related changes in the auditory reaction time of healthy elderly person while walking. *Journal of Physical Therapy Science, 23*(2), 185–188. https://doi.org/10.1589/jpts.23.185

van Herk, I., Arendzen, J., & Rispens, P. (1998). Ten-metre walk, with or without a turn? *Clinical Rehabilitation, 12*(1), 30–35. https://doi.org/10.1191/026921598667081596

Vaportzis, E., Georgiou-Karistianis, N., & Stout, J. (2013). Dual task performance in normal aging: a comparison of choice reaction time tasks. *PLoS ONE, 8*(3), e60265. https://doi.org/10.1371/journal.pone.0060265

Verghese, J., Annweiler, C., Ayers, E., Barzilai, N., Beauchet, O., Bennett, D. A., Bridenbaugh, S., Buchman, A., Callisaya, M., Camicioli, R., Capistrant, B., Chatterji, S., De Cock, A., Ferrucci, L., Giladi, N., Guralnik, J., Hausdorff, J., Holtzer, R., Kim, K., … Wang, C. (2014). Motoric cognitive risk syndrome. *Neurology*, *83*(8), 718–726. https://doi.org/10.1212/WNL.0000000000000717

Verghese, J., Kuslansky, G., Holtzer, R., Katz, M., Xue, X., Buschke, H., & Pahor, M. (2007). Walking while talking: effect of task prioritization in the elderly. *Archives of Physical Medicine and Rehabilitation*, *88*(1), 50–53. https://doi.org/10.1016/j.apmr.2006.10.007

Verghese, J., Wang, C., Ayers, E., Izzetoglu, M., & Holtzer, R. (2017). Brain activation in high-functioning older adults and falls: Prospective cohort study. *Neurology*, *88*(2), 191–197. https://doi.org/10.1212/WNL.0000000000003421

Verghese, J., Wang, C., Lipton, R., Holtzer, R., & Xue, X. (2007). Quantitative gait dysfunction and risk of cognitive decline and dementia. *Journal of Neurology, Neurosurgery, and Psychiatry*, *78*(9), 929–935. https://doi.org/10.1136/jnnp.2006.106914

Verghese, J., & Xue, X. (2010). Identifying frailty in high functioning older adults with normal mobility. *Age and Ageing*, *39*(3), 382–385. https://doi.org/10.1093/ageing/afp226

Vermeij, A., van Beek, A., Olde Rikkert, M., Claassen, J., & Kessels, R. (2012a). Effects of aging on cerebral oxygenation during working-memory performance: A functional near-infrared spectroscopy study. *PLoS ONE*, *7*(9). https://doi.org/10.1371/journal.pone.0046210

Vermeij, A., van Beek, A., Olde Rikkert, M., Claassen, J., & Kessels, R. (2012b). Effects of aging on cerebral oxygenation during working-memory performance: A functional near-infrared spectroscopy study. *PLoS ONE*, *7*(9), e46210. https://doi.org/10.1371/journal.pone.0046210

Vermeij, A., van Beek, A., Reijs, B., Claassen, J., & Kessels, R. (2014). An exploratory study of the effects of spatial working-memory load on prefrontal activation in low- and high-performing elderly. *Frontiers in Aging Neuroscience, 6.* https://doi.org/10.3389/fnagi.2014.00303

Wechsler, D. (1981). *Wechsler adult intelligence scale-revised (WAIS-R).* Psychological Corporation.

West, R. (1996). An application of prefrontal cortex function theory to cognitive aging. *Psychological Bulletin, 120*(2), 272–292. https://doi.org/10.1037//0033-2909.120.2.272

Wollesen, B., Schulz, S., Seydell, L., & Delbaere, K. (2017). Does dual task training improve walking performance of older adults with concern of falling? *BMC Geriatrics, 17*(1), 213. https://doi.org/10.1186/s12877-017-0610-5

Wollesen, B., & Voelcker-Rehage, C. (2014). Training effects on motor–cognitive dual-task performance in older adults: A systematic review. *European Review of Aging and Physical Activity, 11*(1), 5–24. https://doi.org/10.1007/s11556-013-0122-z

Woollacott, M., & Shumway-Cook, A. (2002). Attention and the control of posture and gait: A review of an emerging area of research. *Gait & Posture, 16*(1), 1–14. https://doi.org/10.1016/S0966-6362(01)00156-4

Wu, T., & Hallett, M. (2005). The influence of normal human ageing on automatic movements. *The Journal of Physiology, 562*(Pt 2), 605–615. https://doi.org/10.1113/jphysiol.2004.076042

Wu, T., Kansaku, K., & Hallett, M. (2004). How self-initiated memorized movements become automatic: A functional MRI study. *Journal of Neurophysiology, 91*(4), 1690–1698. https://doi.org/10.1152/jn.01052.2003

Wulf, G. (2013). Attentional focus and motor learning: A review of 15 years. *International Review of Sport and Exercise Psychology, 6*(1), 77–104. https://doi.org/10.1080/1750984X.2012.723728

Wulf, G., Shea, C., & Park, J. (2001). Attention and motor performance: Preferences for and advantages of an external focus. *Research Quarterly for Exercise and Sport, 72*(4), 335–344. https://doi.org/10.1080/02701367.2001.10608970

Yesavage, J., & Sheikh, A. (1986). Geriatric depression scale (GDS). *Clinical Gerontologist, 5*(1–2), 165–173. https://doi.org/10.1300/J018v05n01_09

Yogev, G., Hausdorff, J., & Giladi, N. (2008). The role of executive function and attention in gait. *Movement Disorders : Official Journal of the Movement Disorder Society, 23*(3), 329–472. https://doi.org/10.1002/mds.21720

Yogev-Seligmann, G., Hausdorff, J., & Giladi, N. (2012). Do we always prioritize balance when walking? Towards an integrated model of task prioritization. *Movement Disorders, 27*(6), 765–770. https://doi.org/10.1002/mds.24963

Yogev-Seligmann, G., Rotem-Galili, Y., Mirelman, A., Dickstein, R., Giladi, N., & Hausdorff, J. (2010). How does explicit prioritization alter walking during dual-task performance? effects of age and sex on gait speed and variability. *Physical Therapy, 90*(2), 177–186. https://doi.org/10.2522/ptj.20090043

APPENDIX

Appendix A - Phone Screening

1. In the last year have you participated in a cognitive study? (If yes, at least 6 months should have passed between the ending of the previous study and this study)
2. What is your mother tongue?
3. What is your age?
4. What is your date of birth?
5. What is your highest level of education? (Degree and number of years)
6. Are you right or left-handed?
7. In the last 6 months did you have surgery (or another medical intervention) that required general anesthesia? (If yes, write the date of the anesthesia and we will see them after the delay)
8. Do you have vision or hearing problems that have not been corrected or an abnormal loss for your age?
9. Have you ever had frequent or requiring migraines?
10. Have you ever had neurological or psychiatric problems? If so which ones?
11. Have you ever had a head injury? When?
12. Have you ever lost consciousness?
13. Have you had tremors or involuntary movements?
14. Have you ever had motor problems (walking or manipulating objects) (If severe, exclusion)?
15. Have you ever had dizziness or problems with maintaining your balance (If yes, ask more questions: of what nature, when, frequency, etc.)?
16. Have you ever had an injury to your lower extremities (hip, knee, ankle)?
17. Do you have a cardiovascular condition?
18. Do you have high blood pressure?
19. Do you have diabetes?
20. Do you have arthritis?
21. Do you have epilepsy?
22. Do you have a thyroid condition?
23. Have you ever fallen while walking?
24. What is your weekly consumption of alcohol?
25. Have you ever had alcohol or drug abuse problems?
26. Do you smoke?
27. Do you have problems with concentration or attention?
28. Do you have problems with your memory?
29. Do you have problems finding your words while talking?
30. Have the people close to you noticed a significant difference in these three areas: attention, memory, trouble finding words?
31. Are you physically active regularly (more than 2 times per week) besides activities of daily living? What are these activities?
32. Do you take medication? If yes, do these medications make you feel drowsy or affect your mental state? (List the medications, name and dosage, that are taken regularly).

Appendix B - Montreal Cognitive Assessment (MoCA)

MONTREAL COGNITIVE ASSESSMENT (MOCA)
Version 7.1 Original Version

NAME :
Education :　　　　　Date of birth :
Sex :　　　　　DATE :

VISUOSPATIAL / EXECUTIVE		POINTS

Copy cube

Draw CLOCK (Ten past eleven) (3 points)

[ ]　　　　[ ]　　　　[ ]　[ ]　[ ]
　　　　　　　　　　Contour　Numbers　Hands

__/5

NAMING

[ ]　　　　[ ]　　　　[ ]　　__/3

MEMORY	Read list of words, subject must repeat them. Do 2 trials, even if 1st trial is successful. Do a recall after 5 minutes.		FACE	VELVET	CHURCH	DAISY	RED	No points
		1st trial						
		2nd trial						

ATTENTION	Read list of digits (1 digit/ sec.).	Subject has to repeat them in the forward order	[ ]　2 1 8 5 4	
		Subject has to repeat them in the backward order	[ ]　7 4 2	__/2

Read list of letters. The subject must tap with his hand at each letter A. No points if ≥ 2 errors
[ ]　F B A C M N A A J K L B A F A K D E A A A J A M O F A A B　__/1

Serial 7 subtraction starting at 100	[ ] 93	[ ] 86	[ ] 79	[ ] 72	[ ] 65	
	4 or 5 correct subtractions: **3 pts**, 2 or 3 correct: **2 pts**, 1 correct: **1 pt**, 0 correct: **0 pt**					__/3

LANGUAGE	Repeat : I only know that John is the one to help today. [ ] The cat always hid under the couch when dogs were in the room. [ ]	__/2
	Fluency / Name maximum number of words in one minute that begin with the letter F　[ ] _____ (N ≥ 11 words)	__/1

ABSTRACTION	Similarity between e.g. banana - orange = fruit　[ ] train – bicycle　[ ] watch - ruler	__/2

DELAYED RECALL	Has to recall words WITH NO CUE	FACE [ ]	VELVET [ ]	CHURCH [ ]	DAISY [ ]	RED [ ]	Points for UNCUED recall only	__/5
Optional	Category cue							
	Multiple choice cue							

ORIENTATION	[ ] Date　[ ] Month　[ ] Year　[ ] Day　[ ] Place　[ ] City	__/6

© Z.Nasreddine MD　　**www.mocatest.org**　　Normal ≥ 26 / 30　| TOTAL __/30

Administered by: _______________________________________

Add 1 point if ≤ 12 yr edu

Appendix C - Digit Forward and Digit Backward

<table>
<tr><td>8- MÉMOIRE DE CHIFFRES</td><td>Rythme : 1 chiffre par seconde
Stop : 2 essais d'une même série échoués</td></tr>
</table>

À l'endroit

Question	Réponse	Question	Réponse	Point
1) 1-7		6-3		2 1 0
2) 5-8-2		6-9-4		2 1 0
3) 6-4-3-9		7-2-8-6		2 1 0
4) 4-2-7-3-1		7-5-8-3-6		2 1 0
5) 6-1-9-4-7-3		3-9-2-4-8-7		2 1 0
6) 5-9-1-7-4-2-8		4-1-7-9-3-8-6		2 1 0
7) 5-8-1-9-2-6-4-7		3-8-2-9-5-1-7-4		2 1 0
8) 2-7-5-8-6-2-5-8-4		7-1-3-9-4-2-5-6-8		2 1 0

À rebours

Question	Réponse	Question	Réponse	Point
1) 2-4 (4-2)		5-8 (8-5)		2 1 0
2) 6-2-9 (9-2-6)		7-1-9 (9-1-7)		2 1 0
3) 3-2-7-9 (9-7-2-3)		4-9-6-8 (8-6-9-4)		2 1 0
4) 1-5-2-8-6 (6-8-2-5-1)		6-1-8-4-3 (3-4-8-1-6)		2 1 0
5) 5-3-9-4-1-8 (8-1-4-9-3-5)		7-2-4-8-5-6 (6-5-8-4-2-7)		2 1 0
6) 8-1-2-9-3-6-5 (5-6-3-9-2-1-8)		4-7-3-9-1-2-8 (8-2-1-9-3-7-4)		2 1 0
7) 9-4-3-7-6-2-5-8 (8-5-2-6-7-3-4-9)		7-2-8-1-9-6-5-3 (3-5-6-9-1-8-2-7)		2 1 0

Appendix D - Digit Symbol Substitution Test (DSST)

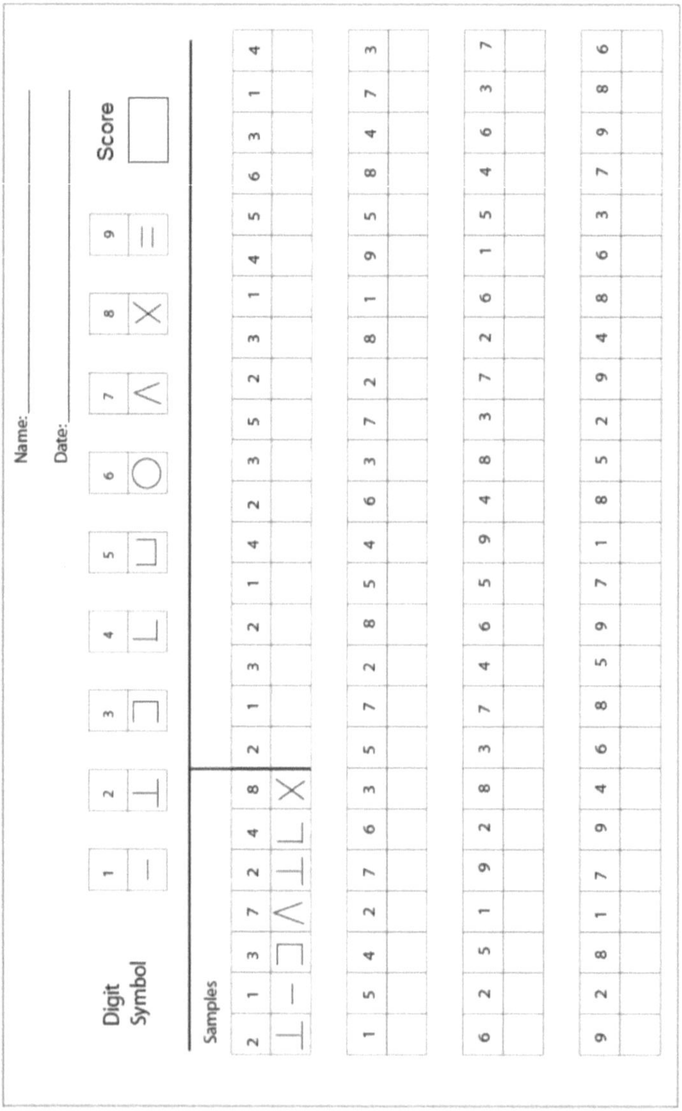

Figure 13.1 The Digit Symbol Substitution Test (DSST). The athlete completes as many boxes as possible in 90 seconds

Appendix E - Trail Making Test (TMT) Parts A and B

Trail Making

part A

EXEMPLE

NOM_________________________________

Date_________________________________

Temps__________   Erreur____________

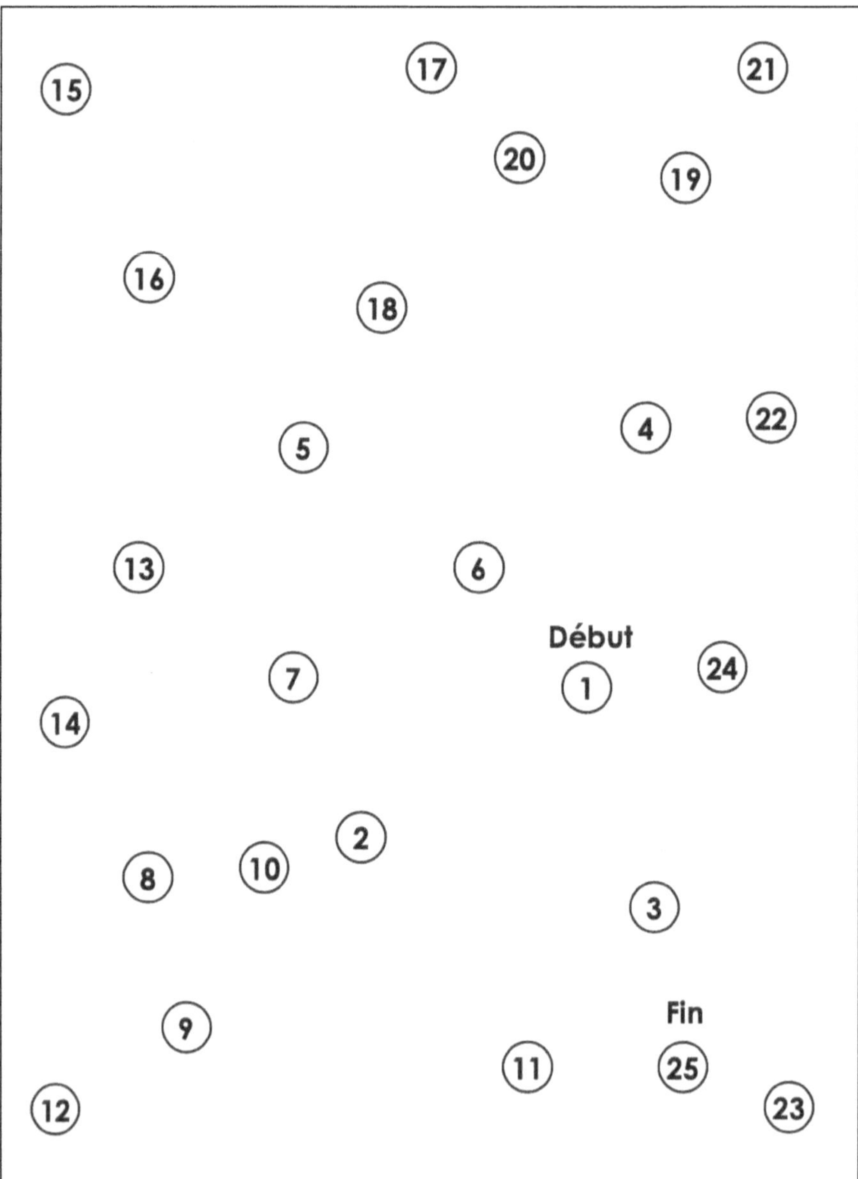

15
17
21
20
19
16
18
4
22
5
13
6
Début
7
1
24
14
2
8
10
3
9
Fin
11
25
12
23

Trail Making

part B

EXEMPLE

NOM_______________________________

Date_______________________________

Temps__________   Erreur___________

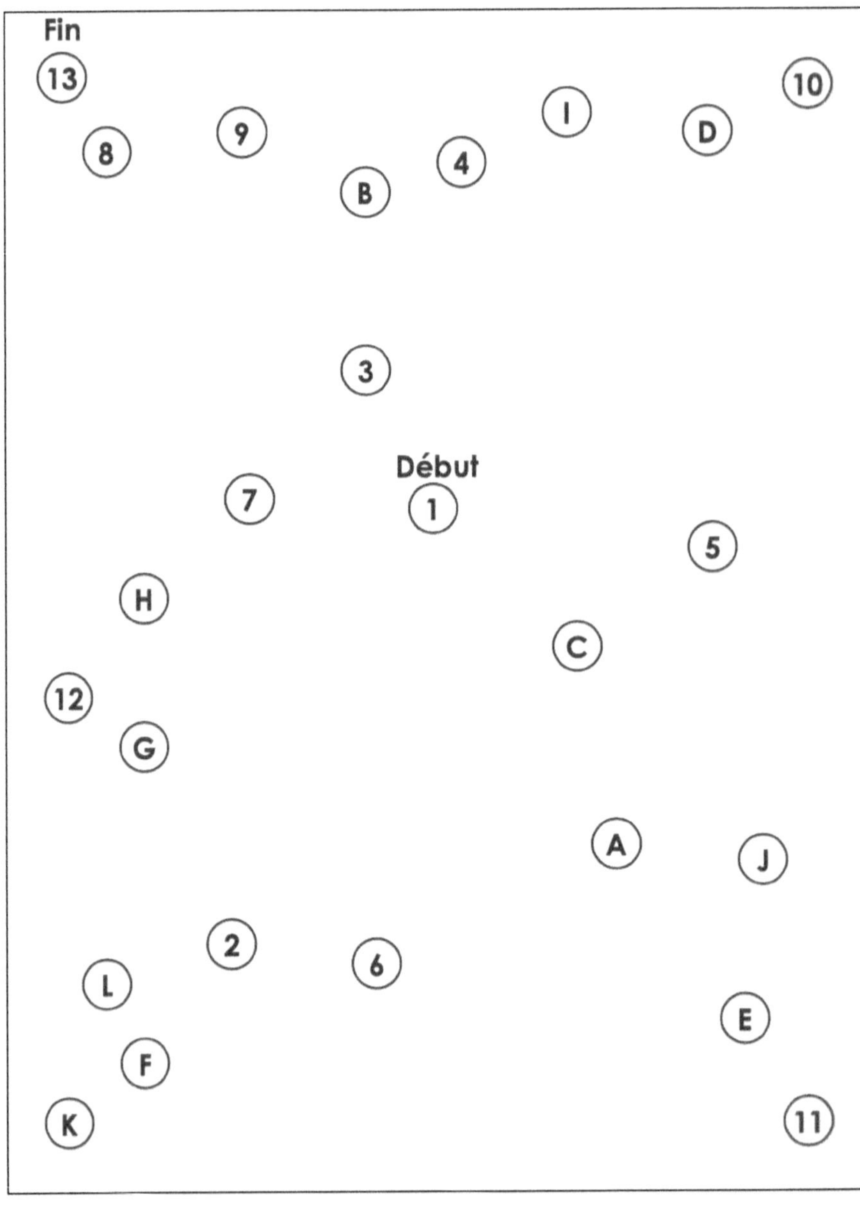

Fin
13
10
8
9
I
D
B
4
3
Début
7
1
5
H
C
12
G
A
J
2
6
L
E
F
K
11

Appendix F - Short Physical Performance Battery (SPPB)

1. Repeated Chair Stands
Instructions: Do you think it is safe for you to try and stand up from a chair five times without using your arms? Please stand up straight as quickly as you can five times, without stopping in between. After standing up each time, sit down and then stand up again. Keep your arms folded across your chest. Please watch while I demonstrate. I'll be timing you with a stopwatch. Are you ready? Begin Grading: Begin stopwatch when subject begins to stand up. Count aloud each time subject arises. Stop the stopwatch when subject has straightened up completely for the fifth time.
Also stop if the subject uses arms, or after 1 minute, if subject has not completed rises, and if concerned about the subject's safety. Record the number of seconds and the presence of imbalance. Then complete ordinal scoring.
Time: _____sec (if five stands are completed)
Number of Stands Completed: 1 2 3 4 5
Chair Stand Ordinal Score: _____
0 = unable
1 = > 16.7 sec
2 = 16.6-13.7 sec
3 = 13.6-11.2 sec
4 = < 11.1 sec

2. Balance Testing
Begin with a semi-tandem stand (heel of one foot placed by the big toe of the other foot). Individuals unable to hold this position should try the side-by-side position. Those able to stand in the semi-tandem position should be tested in the full tandem position. Once you have completed time measures, complete ordinal scoring.

a. Semi-tandem Stand
Instructions: Now I want you to try to stand with the side of the heel of one foot touching the big toe of the other foot for about 10 seconds. You may put either foot in front, whichever is more comfortable for you. Please watch while I demonstrate.
Grading: Stand next to the participant to help him or her into semi-tandem position. Allow participant to hold onto your arms to get balance. Begin timing when participant has the feet in position and let's go.
Circle one number
2. Held for 10 sec
1. Held for less than 10 sec; number of seconds held _____
0. Not attempted

b. Side-by-Side stand
Instructions: I want you to try to stand with your feet together, side by side, for about 10 sec. Please watch while I demonstrate. You may use your arms, bend your knees, or move your body to maintain your balance, but try not to move your feet. Try to hold this position until I tell you to stop.

Grading: Stand next to the participant to help him or her into the side-by-side position. Allow participant to hold onto your arms to get balance. Begin timing when participant has feet together and let's go.
Grading
2. Held of 10 sec
1. Held for less than 10 sec; number of seconds held______
0. Not attempted

c. Tandem Stand
Instructions: Now I want you to try to stand with the heel of one foot in front of and touching the toes of the other foot for 10 sec. You may put either foot in front, whichever is more comfortable for you. Please watch while I demonstrate.
Grading: Stand next to the participant to help him or her into the side-by-side position. Allow participant to hold onto your arms to get balance. Begin timing when participant has feet together and let's go.
Grading
2. Held of 10 sec
1. Held for less than 10 sec; number of seconds held______
0. Not attempted

Balance Ordinal Score: ______
0 = side by side 0-9 sec or unable
1 = side by side 10, <10 sec semitandem
2 = semitandem 10 sec, tandem 0-2 sec
3 = semitandem 10 sec, tandem 3-9 sec
4 = tandem 10 sec

3. 8' Walk (2.44 meters)
Instructions: This is our walking course. If you use a cane or other walking aid when walking outside your home, please use it for this test. I want you to walk at your usual pace to the other end of this course (a distance of 8'). Walk all the way past the other end of the tape before you stop. I will walk with you. Are you ready?
Grading: Press the start button to start the stopwatch as the participant begins walking.
Measure the time take to walk 8'. Then complete ordinal scoring.
Time: ______ sec
Gait Ordinal Score: ______
0 = could not do
1 = >5.7 sec (<0.43 m/sec)
2 = 4.1-6.5 sec (0.44-0.60 m/sec)
3 = 3.2-4.0 (0.61-0.77 m/sec)
4 = <3.1 sec (>0.78 m/sec)
Summary Ordinal Score: ______
Range: 0 (worst performance) to 12 (best performance). Shown to have predictive validity showing a gradient of risk for mortality, nursing home admission, and disability.

Appendix G - Falls Efficacy Scale – International (FES-I)

Now we would like to ask some questions about how concerned you are about the possibility of falling. For each of the following activities, please **circle** the opinion closest to your own to show how concerned you are that you might fall if you did this activity. Please reply thinking about how you usually do the activity. If you currently don't do the activity (e.g. if someone does your shopping for you), please answer to show whether you think you would be concerned about falling IF you did the activity.

		Not at all concerned	*Somewhat concerned*	*Fairly concerned*	*Very concerned*
1	Cleaning the house (e.g. sweep, vacuum or dust)	*1*	*2*	*3*	*4*
2	Getting dressed or undressed	*1*	*2*	*3*	*4*
3	Preparing simple meals	*1*	*2*	*3*	*4*
4	Taking a bath or shower	*1*	*2*	*3*	*4*
5	Going to the shop	*1*	*2*	*3*	*4*
6	Getting in or out of a chair	*1*	*2*	*3*	*4*
7	Going up or down stairs	*1*	*2*	*3*	*4*
8	Walking around in the neighbourhood	*1*	*2*	*3*	*4*
9	Reaching for something above your head or on the ground	*1*	*2*	*3*	*4*
10	Going to answer the telephone before it stops ringing	*1*	*2*	*3*	*4*
11	Walking on a slippery surface (e.g. wet or icy)	*1*	*2*	*3*	*4*
12	Visiting a friend or relative	*1*	*2*	*3*	*4*
13	Walking in a place with crowds	*1*	*2*	*3*	*4*
14	Walking on an uneven surface (e.g. rocky ground, poorly maintained pavement)	*1*	*2*	*3*	*4*
15	Walking up or down a slope	*1*	*2*	*3*	*4*
16	Going out to a social event (e.g. religious service, family gathering or club meeting)	*1*	*2*	*3*	*4*

Appendix H - Geriatric Depression Scale (GDS)

Geriatric Depression Scale

1.	Are you basically satisfied with your life?	NO	yes
2.	Have you dropped many of your activities and interests?	YES	no
3.	Do you feel that your life is empty?	YES	no
4.	Do you often get bored?	YES	no
5.	Are you hopeful about the future?	NO	yes
6.	Are you bothered by thoughts that you just cannot get out of your head?	YES	no
7.	Are you in good spirits most of the time?	NO	yes
8.	Are you afraid that something bad is going to happen to you?	YES	no
9.	Do you feel happy most of the time?	NO	yes
10.	Do you often feel helpless?	YES	no
11.	Do you often get restless and fidgety?	YES	no
12.	Do you prefer to stay home at night, rather than go out and do new things?	YES	no
13.	Do you frequently worry about the future?	YES	no
14.	Do you feel that you have more problems with memory than most?	YES	no
15.	Do you think it is wonderful to be alive now?	NO	yes
16.	Do you often feel downhearted and blue?	YES	no
17.	Do you feel pretty worthless the way you are now?	YES	no
18.	Do you worry a lot about the past?	YES	no
19.	Do you find life very exciting?	NO	yes
20.	Is it hard for you to get started on new projects?	YES	no
21.	Do you feel full of energy?	NO	yes
22.	Do you feel that your situation is hopeless	YES	no
23.	Do you think that most people are better off than you are?	YES	no
24.	Do you frequently get upset over little things?	YES	no
25.	Do you frequently feel like crying?	YES	no
26.	Do you have trouble concentrating?	YES	no
27.	Do you enjoy getting up in the morning?	NO	yes
28.	Do you prefer to avoid social gatherings?	YES	no
29.	Is it easy for you to make decisions?	NO	yes
30.	Is your mind as clear as it used to be?	NO	yes